PREPARATION AND PROPERTIES OF MONOMERS, POLYMERS AND COMPOSITE MATERIALS

PREPARATION AND PROPERTIES OF MONOMERS, POLYMERS AND COMPOSITE MATERIALS

ANTONIO BALLADA
AND
GENNADY E. ZAIKOV
EDITORS

Nova Science Publishers, Inc.
New York

For permission to use material from this book please contact us:
Telephone 631-231-7269; Fax 631-231-8175
Web Site: http://www.novapublishers.com

LIBRARY OF CONGRESS CATALOGING-IN-PUBLICATION DATA

Preparation and properties of monomers, polymers, and composite materials / Antonio Ballada and Gennady E. Zaikov (editors).

p. cm.

Includes index.

ISBN-13: 978-1-60021-557-5 (hardcover)

ISBN-10: 1-60021-557-2 (hardcover)

1. Monomers. 2. Polymers. 3. Composite materials. I. Ballada, Antonio. II. Zaikov, Gennadii Efremovich.

TP247.P67 2006

668.9--dc22 2006103351

Published by Nova Science Publishers, Inc. ✦ New York

CONTENTS

"Nothing to be compared on speed to an idea"
Cicero, Ancient Rome

"There is nothing further, than Yesterday.
There is nothing more close, than Tomorrow"
The Chinese proverb of Confucius time

PREFACE

One of the proverbs that is written in an epigraph, reflects very well a situation in science. A science (and scientists) should look forward (certainly, except for history and disciplines accompanying it). We believe that the given book looks forward, and the results presented in it will cause doubtless interest in the broad audience of readers and the scientists working in the field of Pure and Applied Chemistry. Hereafter we inscribe two educational stories:

A story about two mice. Two mice have gotten in two vessels that are half filled with milk. The first mouse has realized that there is no way out. It gives up its fight for life and has sunk, and the Lord has taken its soul. The second mouse has also realized that it is impossible to get out of this vessel, but began to fight, trying to prolong its life. Marvelously, it whipped the milk into butter, climbed up over a piece of a butter and escaped from the vessel. Conclusions: one has to fight to the end even in the case when the position is completely hopeless.

A story about a sparrow. There was a strong frost in Siberia (minus 50 degrees Celsius) and the sparrow got frozen while it was flying. It got benumbed, has fallen on a road and has already said goodbye to its life. Meanwhile a cow went by and made droppings on the sparrow. The sparrow warmed up in the warm droppings, revived and started to chirp. A cat went by. It heard chirping, ran up and ate the sparrow. Conclusions: 1). Stay indoor in case of harsh weather, otherwise you can put your health at risk; 2). If someone messed you up with excrements it does not automatically mean, that it is very bad; 3). If you have are in excrements - do not chirp.

These stories are not directly related to the given compilation book, but there is a certain indirect link (as a way of survival of Science and Scientists, in particular, in Russia).

Dr. Antonio Ballada
Rome, Italy
Prof. Gennady Zaikov
Moscow, Russia

In: Preparation and Properties of Monomers, Polymers... ISBN 978-1-60021-557-5
Editors: A. Ballada and G. E. Zaikov, pp. 1-11

Chapter 1

ENHANCEMENT OF MISCIBILITY IN MULTICOMPONENT SOLUTIONS ON THE BASIS OF THREE POLYMERS AND COMMON SOLVENTS

Yu. P. Miroshnikov* and V. N. Kuleznev
M. V. Lomonosov Moscow State Academy of Fine Chemical Technology,
86, Vernadsky Ave., Moscow 117571, Russia

ABSTRACT

Mutual solubility in ternary systems comprising solutions of three immiscible polymers and common solvents has been studied. Cloud point measurements were performed and the solubility isotherms were plotted as functions of composition in ternary coordinates of Gibbs triangle. For several systems composed of different polymers and different common solvents the effect of miscibility enhancement in ternary systems compared to corresponding binary mixtures has been detected. The first version of the explanation presented is based on experimental data and considerations known elsewhere, namely: phase separation in polymer solutions takes place when a critical size of macromolecular associates is reached; progressive contraction of similar molecular coils proceeds on addition of the third and subsequent immiscible polymers. These mean that any associate of critical size in ternary polymer system may consist of a larger number of macromolecular coils compared to that in binary system. In other words, solubility enhancement between polymers in a common solvent, induced by addition of a third immiscible polymer, is presumably due to the increase of the state of association of polymer components, the critical sizes of the associates are being unchanged. The alternative version of the explanation leading to the same conclusion is the reduction of the partial concentrations of polymer components on the increase of their number in a mixture. Experimental data on the intensity of light scattering in homogeneous solutions of binary and ternary polymer blends, which are in accordance with the proposed mechanisms, are also presented and discussed.

* Corresponding author: Yuri P. Miroshnikov; E-mail miroshnikov@mitht.ru; Fax: +7 (495) 246-4741; Tel: +7 (495) 246-0958.

Keywords: polymer solutions, ternary polymer blends, miscibility, cloud point diagrams, light scattering.

INTRODUCTION

Most part of polymer pairs are immiscible, i.e. on mixing they form heterogeneous systems [1-3]. Such blends exhibit non-additive property-composition relationships making prediction of their behavior doubtful. From the other side, sometimes, significant benefits in properties of immiscible blends compared to individual components can be obtained which makes these system very attractive to science and commercial use.

Interaction between polymer components in blends making strong impact on final properties is investigated usually with the help of phase diagrams plotted in temperature-composition coordinates. Construction of the diagrams in absence of solvents is very time consuming procedure owing to long time of establishing true phase equilibrium. When LST or UST is not experimentally accessible the construction of phase diagrams has little sense at all. In this case a general alternative is the measurement of mutual solubility of polymers at constant temperature. Transition from homogeneous to heterogeneous system on varying composition of a mixture is usually identified by measurements of glass transition temperatures, optical density (cloud point experiments), etc.

For ternary systems comprising three immiscible polymer pairs these data are plotted in the form of isotherms in Gibbs composition triangle. As far as miscibility (mutual solubility) of commonly used polymers is usually within few percent such phase diagrams bear minor information. The reason is the regions of homogeneous compositions is small and localized very close to the angular points of the triangle.

In this sense the use of polymer solutions is more preferable. First, the dilution of a blend with a common solvent allows one to select proper concentration of blend solution and thus to increase the accuracy of cloud point experiments. Second, phase equilibrium in diluted systems is readily obtainable. And, finally, in contrast to bulk blends in which alternative phases interact only through the interfaces the interaction of polymers in solutions is implemented on the level of coils or their associates. This fundamentally different manner of polymer-polymer interactions in diluted and bulk systems does not allow quantitative extrapolation of the miscibility in solutions to that in solid state. Nevertheless, multicomponent polymer solutions are widely used in different industrial technologies [1, 3]. Therefore a deeper understanding of the interactions and solubility phenomena in such systems is still relevant.

Studies on the effect of the addition of a third component on mutual solubility of polymers in ternary systems are the subjects of a number of publications [1, 2, 4-12]. It is known that the chemical nature of a solvent and other low-molecular weight additives play an important role in the interactions between polymers. More than 20 low molecular weight additives to binary blend solutions were studied [5] showing both the increase or decrease of polymer-polymer solubility. Results obtained are explained in terms of the selective solvation of polymer coils by the molecules of solvents which effects in its turn the state of association of similar macromolecules. Additives which increase or decrease the state of association of polymer coils resulted in reduction or enhancement of polymer-polymer solubility respectively. Similar results were obtained in [6].

Systems including more than two immiscible polymers represent complicate picture of different types of interactions [8, 9-12]. Kuhn et al. [12] studied light scattering in solutions comprising three and four immiscible polymers in a common solvent. They showed that the miscibility of polymers increased progressively on addition of the third and fourth polymer.

The aim of this work is further investigation of the effect of addition of third polymer on mutual solubility of polymers in solutions.

EXPERIMENTAL

Materials

Polymers used in this study and some of their characteristics are listed in Table. Cis-1,4-polybutadiene BR-305 (PB) and cis-1,4-polyisoprene IR-305 (PI) were produced by 'Shell'. Commercial grades of atactic block-polymerized polystyrene (PS), rigid poly(vinyl chloride) S-70 (PVC), cis-1,4-polybutadiene SKD, polyisobutylene P-85 (PIB), styrene butadiene copolymer SKMS-30 ARK (30.0 wt. % of styrene; SBR), butadien acrylonitrile rubbers, containing 18.0 (NBR-18) and 40.0 (NBR-40) wt. % of AN and poly(methyl vinyl siloxane) SKTV-1B (SR) were purchased from different domestic suppliers. All polymers were purified by double precipitation of their solutions in excess of methanol. Toluene solutions (5.0 g/dl) of SR were purified by centrifugation for 1 hour at 7000 rpm. Benzene, toluene and CCl_4 employed as common solvents were distilled before use. To prevent oxidation all solutions were stabilized with 0.1 wt. % of colorless "Ionol" antioxidant. Refractive indices of polymers and solvents were borrowed from literature [13].

Cloud Point Measurements

A technique of cloud point measurements at 20 ^{o}C is illustrated by Figure 1. This picture explains the most common case in this study when at used general concentration of a ternary solution one binary polymer blend B/C is homogeneous (i.e. general concentration of the solution is lower compared to that at which phase separation in this blend takes place) while the other two blends A/B and A/C are heterogeneous. Solution A is added to solution B (or C) by small portions until phase separation between polymers A and B (or C) takes place. Then solution A is added to homogeneous blend solutions A/C 75/25, 50/50 and 25/75. It is clear that ternary compositions determined by the cloud point measurements lay on the lines connecting the apex of a triangle with corresponding binary compositions shown on its lower side. The experimental points obtained are connected by an isotherm line separating homogeneous and heterogeneous compositions. The same procedure is used when solutions of polymers B and C and their mixtures are added to solution A.

These experiments were made in glass test tubes supplied with grounded stoppers to avoid solvent evaporation. Phase separation in the mixtures was identified visually in transmitted red light. This technique was shown to have advantages compared to instrumental methods (i.e. nephelometry), especially when refractive indices of mixed polymers did not differ very much.

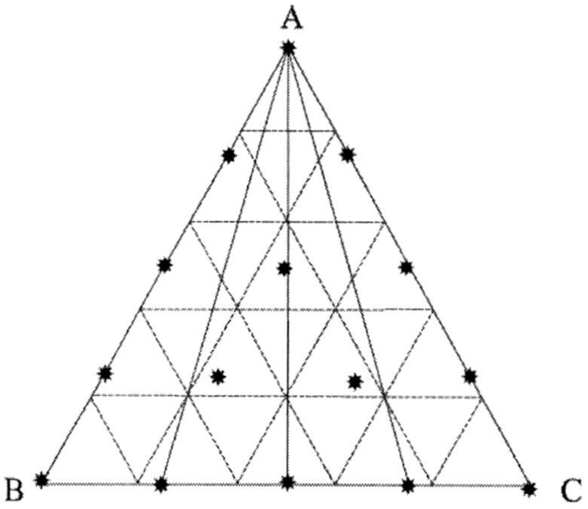

Figure 1. Schematic representation of the cloud point measurement technique and the lattice points in which light scattering have to be measured according to forth-order polynomial model.

Light Scattering

Light scattering (more precisely, excessive light scattering, R_{90}) of homogeneous polymer solutions at right angles and the wave length 546 nm of non-polarized light was measured on the differential photoelectric nephelometer FPS-2 at 20 °C.

Triangle diagrams R_{90}–composition were computed on the basis of simplex lattice design proposed by Scheffe [12]. The fact of the matter is the development of a mathematical model of a response surface describing the property-composition relationship in Gibbs triangle coordinates. Values of 15 measurements of R_{90} at different 15 compositions (Figure 1) were needed to build a 4-order polynomial used normally in this work. Additional 2-3 measurements at arbitrary ternary compositions were made to check the validity of a model.

Results and Discussion

Cloud Point Curves

As the first step, concentration of solution, C*, corresponding to phase separation in each binary combination of polymers at fixed composition ratio 50/50 has been determined. On the basis of these data general concentrations, C_g, of solutions of ternary polymer blends have been chosen. General concentration was selected in a way providing that: 1) one polymer pair is homogeneous while another two pairs are heterogeneous; 2) all three binary blends are heterogeneous.

Figure 2 shows the cloud point curves for ternary blend system PIB/PB/PS with CCl_4 as a common solvent at C_g = 4.0 g/dl. At this general concentration of solutions a pair PB/PS is homogeneous in entire composition range while mutual solubility in the rest two polymer pairs is significantly lower, so that they are heterogeneous.

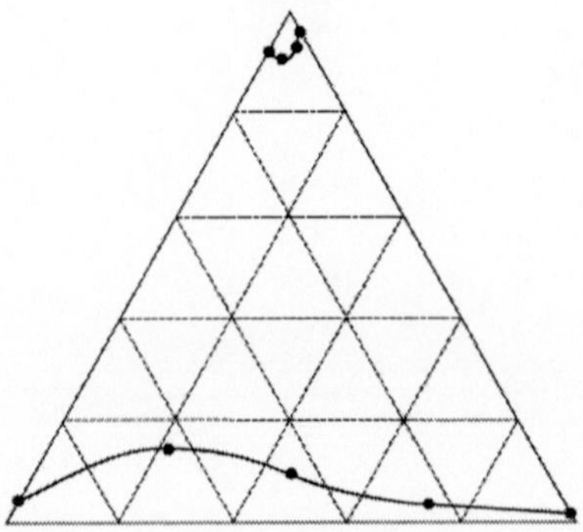

Figure 2. Cloud point diagram for ternary polymer blends PIB/PB/PS. General concentration of polymers in CCl_4 is $C_g = 4.0$ g/dl.

Note that the term "solubility of polymers" used through this work must be treated as a solubility of equally concentrated solutions of corresponding polymers. On addition of the solution of PIB to the solutions of PB, PS or their mixtures or when solutions of the later are added to PIB solution, critical compositions, φ_{crit}, are reached beyond which phase separation takes place. For example solubilities of PIB in PS and in PB are both about $\varphi_{crit} = 1.4$ vol. %; while these of PB and PS in PIB are 9.9 and 2.4 % respectively. At the same time solubility of PIB in a mixture PB/PS 75/25 is increased up to 13 vol. %. Convex shape of the isotherm means apparently that addition of the third polymer results in growth of solubility in ternary system. Similar conclusion is evident when solutions of PB, PS and their mixtures are added to the solution of PIB. Concave shape of this isotherm indicates that solubility of binary blends PB/PS in PIB are higher compared to that of individual PB and PS.

The next ternary system represented by Figure 3a differs from the previous one by substitution of PB by PI. Two general concentrations $C_g = 2.5$ g/dl (curves 1, 4) and $C_g = 2.25$ g/dl (curves 2, 3) were used. Reduction of C_g results in progressive increase of the range of homogeneous ternary compositions. One can easily predict that further decrease of general concentration of solutions may lead to the formation of "sand-glass" type diagram. It is interesting to note that the reduction of C_g from 2.5 to 2.25 g/dl results in minor increase of solubility of PIB in PI and PS while solubility of both PI and PS in PIB increases notably. Besides, solubility in ternary blends seems more sensitive to the variation of C_g than that in binary ones.

If general concentration in the same system is increased to $C_g = 4.00$ g/dl all three polymer pairs undergo phase separation. In this case, illustrated by Figure 3b, the cloud point curves are located close to the summits (tops) of the triangle where the compositions are enriched by one of the pure component and the regions of the one-phase compositions both in binary and ternary mixtures become very small. Therefore, these areas of the triangle diagram were enlarged in Figure 3b for clarity. All three cloud point curves are convex towards ternary compositions indicating enhanced solubility in ternary blends compared to that in corresponding binary mixtures.

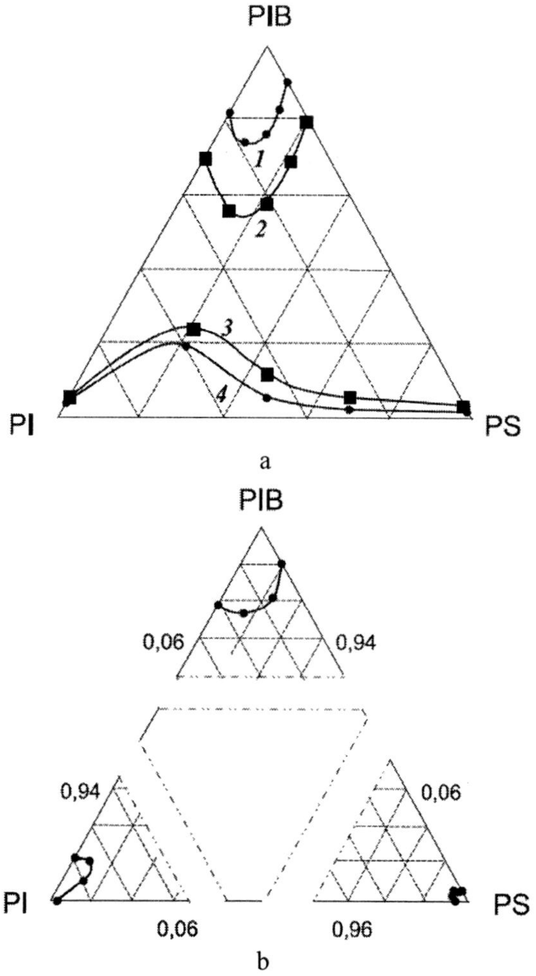

Figure 3. Cloud point curves for PIB/PI/PS system. General concentrations of carbon tetrachloride solutions are: a) 2.5 g/dl (curves 1 and 4) and 2.3 g/dl (curves 2 and 3); b) 4.0 g/dl.

A different type of cloud point diagram for system PS/SR/PI is shown in Figure 4. Values of C* for binary mixtures PS/SR and PS/PI in toluene are similar (~ 2.25 g/dl) while that for a blend SR/PI is noticeably lower (~1.80 g/dl). Therefore, ternary mixtures can be obtained whether in the form of completely heterogeneous solution systems if $C_g > 2.25$ g/dl or in the form of two homogeneous and one heterogeneous mixtures provided general concentration of the solution is between 2.25 and 1.80 g/dl. The first case is described by a diagram similar to that shown in Figure 3b. The second case is shown in Figure 4 where a cloud point isotherm encloses heterogeneous mixtures.

Polymer components in the diagram presented in Figure 5 are PB and its random copolymers NBR-18 and SBR. At general concentration of 5.0 g/dl in benzene a PB/SBR polymer pair forms homogeneous solutions while another two pairs are heterogeneous. Despite the fact that the common solvent and two of polymers have been changed, this diagram is qualitatively similar to the previous ones shown in Figures 1-3.

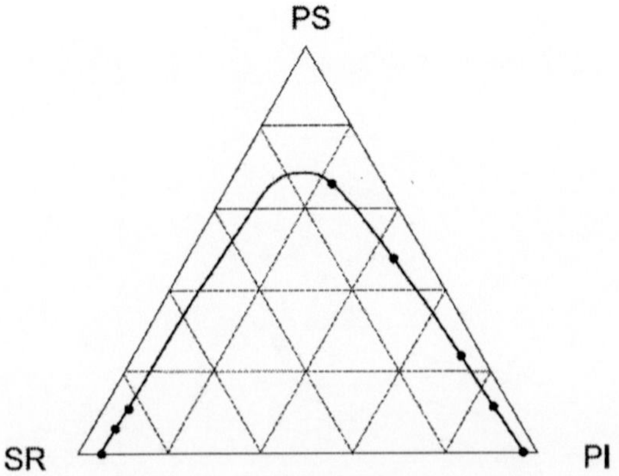

Figure 4. Cloud point diagram for ternary system PS/SR/PI. At general concentration of polymers in toluene 2.0 g/dl mixtures PS/SR and PI/PS are both homogeneous while mixture SR/PI is heterogeneous.

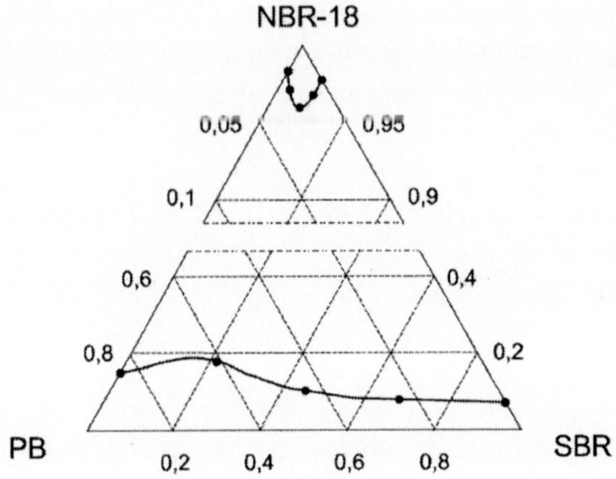

Figure 5. Ternary cloud point diagram for elastomeric system BNR-18/PB/SBR. General concentration of polymers in benzene is 5.0 g/dl.

In contrast to the systems presented before in which all three polymer pairs are immiscible, the ternary system shown in Figure 6 comprises one miscible pair. It was proved elsewhere [1, 2, 15] that owing to specific interactions NBR-40 and PVC are miscible in entire composition range even in absence of a solvent. Nevertheless, this diagram does not differ in principle from these discussed earlier. This leads to a conclusion that ternary blends comprising one homogeneous and two heterogeneous pairs have similar shapes of the cloud point curves irrespectively on whether miscible or immiscible polymers form the homogeneous solution.

The most part of cloud point curves discussed herein are asymmetrical. Origination of this effect is seems owing to different solubilities of polymers in binary combinations represented by lateral sides of triangle diagrams. For instance, at general solution concentration used in Figure 5, solubility of NBR-18 in PB is higher compared to that in SBR. Therefore, higher solubility in ternary compositions enriched by PB component takes place. In contrast, when solubility of SKD in PVC and in NBR-40 is similar, as can be seen in Figure 6, the cloud point isotherms are symmetrical.

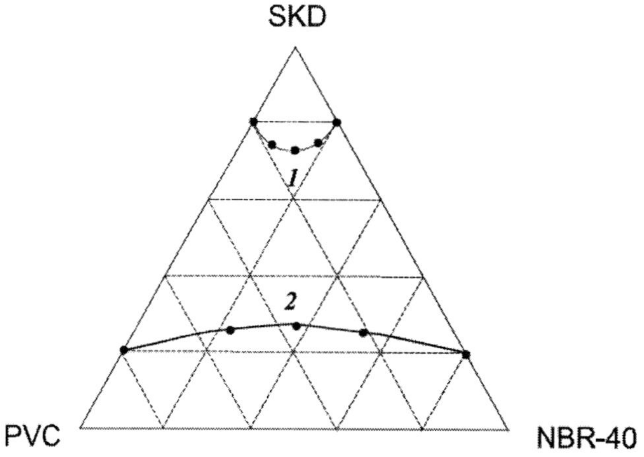

Figure 6. Cloud point diagram for the SKD/PVC/BNR-40 system in which a pair PVC/BNR-40 is completely miscible. General concentration in cyclohexanone is 3.0 g/dl.

It is worth to emphasize that phase separation in immiscible ternary mixtures takes place as a result of interactions between each sort of polymer molecules A-B, B-C and A-C. Critical concentrations of solutions corresponding to phase separation C* for each binary combination are generally different. It was experimentally proved that when initially homogeneous solution of three polymers is gradually concentrated (say, by solvent evaporation) a two-step phase separation takes place. The first step occurs when general concentration of a ternary solution become higher compared to minimal C* in one of the binary combination. This results in the formation of two non-equaled volume layers one of which is a solution of one polymer, and another one is a homogeneous solution of the rest two polymers. Three layers appear on further increase of general concentration. Therefore, one can conclude that a cloud point measured in ternary blends is due to phase separation between two polymers having minimal C*.

Light Scattering

To further clarify the above presented data illustrating enhanced solubility in solutions of three immiscible polymers in a common solvent, a series of experiments on light scattering have been performed. Before discussion of these results, some important features concerning multicomponent polymer solutions are worth to be mentioned. Provided two polymers are immiscible the addition of a solution of one polymer to another one is equivalent to the addition of a non-solvent [1, 2, 4-9]. This results in excessive association of similar molecular coils the extent of which is the higher the lower the polymer-polymer miscibility and/or the closer the concentration of a system to phase separation. When a critical size of associates is reached the phase separation in solution occurs.

Excessive association of similar molecules on addition of a second polymer solution begins long before the moment of phase separation. Therefore, a value of light scattering, R_{90}, of homogeneous binary and ternary polymer solutions which is a function of the size of associates (or generally, fluctuation of concentration) can supply information about a tendency of a system towards phase separation. At the same time one has to remember that

R_{90} is also a strong function of polymer – solvent and polymer – polymer refractive index differences.

Light scattering of homogeneous solutions of the above mentioned systems were measured at 20 °C. General concentration of a ternary system was adjusted a little bit lower compared to the lowest C* in corresponding binary blends. As an example two ternary diagrams R_{90} – composition plotted with the use of simplex lattice design technique [14] are presented in Figure 7.

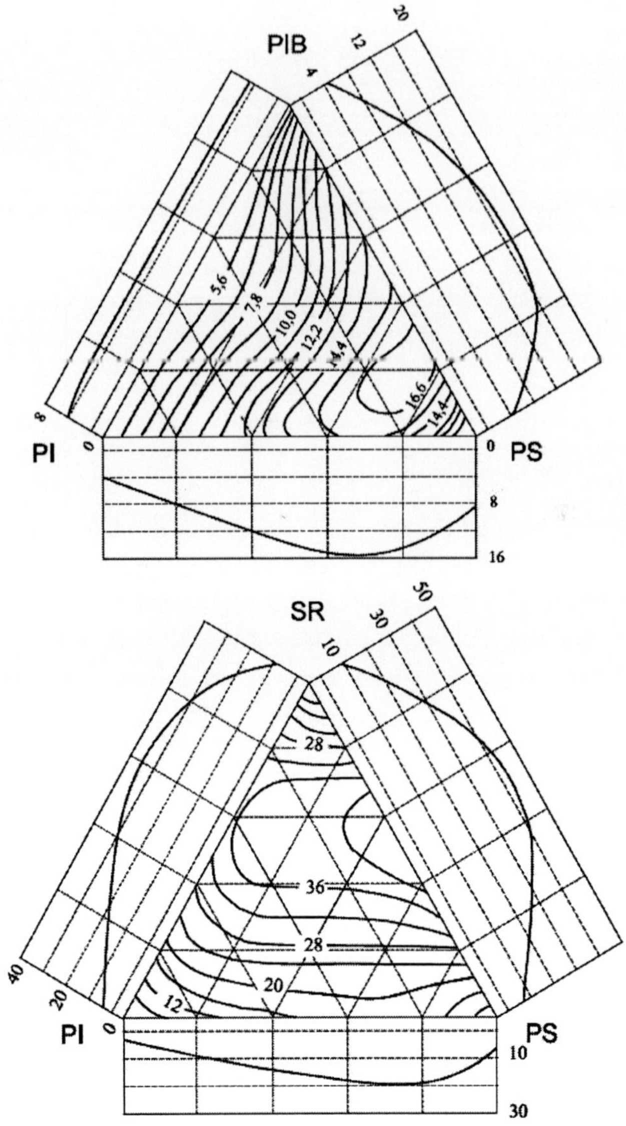

Figure 7. Ternary diagrams R_{90} – composition ($R_{90} \cdot 10^6$ cm^{-1}) for homogeneous solutions: a) system PIB/PI/PS, C_g = 1.8 g/dl; b) system SR/PI/PS, C_g = 1.5 g/dl.

Light scattering diagram for the carbon tetrachloride solutions of PIB/PI/PS system at general concentration of 1.80 g/dl is depicted in Figure 7a. Curves R_{90} – composition for binary blends PI/PS and PIB/PS are characterized by intensive maxima reflecting high extent of association of similar coils [1, 4-6]. Weak maximum of R_{90} for binary blends PIB/PI seems does not mean the absence of excessive association. Instead, this effect is obviously owing to small difference between refractive indices of PIB (~1.509) and PI (~1.519) (see Table). It is important to emphasize that no extreme points of the R_{90} for ternary compositions are detectable. Hence, one can speculate that the sizes of coil associates in ternary mixtures do not exceed those in corresponding binary combinations.

Table. Some characteristics of polymers and solvents

Material	Molecular weight, $M_w \cdot 10^{-5}$ (g/mol)	Refractive index, n_D^{20}
SKD	9.2	1.5156
BR-305 (PB)	1.9	1.5156
IR-305 (PI)	8.5	1.5191
SBR	-	1.5350
SR	5.3	1.4040
PIB	4.8	1.5080
PVC	3,4	1.5440
PS	3.7	1.5935
Benzene		1.5011
Toluene		1.4969
CCl_4		1.4456

Nevertheless, it is essential to compose a ternary blend solution in a way providing each binary polymer mixture demonstrates maxima in light scattering. This case is realized in toluene solutions of SR/PI/PS system with general concentration of 1.50 g/dl shown in Figure 7b. Critical concentrations of phase separation in binary blends SR/PI, PI/PS and PS/SR are 1.6, 1.7 and 2.3 g/dl respectively. Besides, refractive indices of a solvent and constituent polymers differ markedly resulting in significant maxima on each binary R_{90} – composition curve. But again, no maximum light scattering inside a triangle appears.

The data presented means that addition of third polymer to binary polymer solution results in growth of polymer solubility. These are in line with the discussed above results on cloud point measurements.

An explanation of the effect of the increase of polymer solubility in solutions on addition of a third and subsequent immiscible polymers can be established on the following findings known from literature. It was shown [1, 4-8] that phase separation in binary polymer solutions occurs when a critical size of associates is reached. From the other hand, according to the data by Kuhn et al. [12] obtained with dilute solutions an addition of third and subsequent polymers results in progressive contraction of molecular coils. It turns out then, that smaller coils form smaller associates. Hence, a further increase of concentration is needed to reach critical size of such associates and to provoke phase separation in ternary system solution. This means in fact an increase of polymer solubility in multicomponent solutions.

Another possible explanation is the reduction of partial concentration of each polymer component on increasing of their number in a mixture leading to the enhancement of their solubility.

Conclusions

Phase separation in multicomponent polymer blends dissolved in a common solvent is complex phenomena which has not been widely documented and discussed in literature. Therefore the explanations of experimental data presented above can be considered as possible versions which are free for further dispute and criticism. The main key of the first of the explanations of the increase of solubility is (according to Kuhn et al.) the reduction of the sizes of associates formed by the contacted coils in presence of a third polymer component. Therefore, a further increase of concentration of ternary solution is needed to reach a critical size of the associates at which phase separation occurs. This interpretation undermines obviously that the state of association of an associate of critical size in ternary system is higher compared to that in binary mixture. The alternative version of the explanation leading to the same conclusion is the reduction of the partial concentrations of polymer components on the increase of their number in a mixture. In any case direct measurements of density of the associates in binary and multicomponent polymer solutions are desirable to choose from these two concepts or to propose the other ones.

References

[1] Kuleznev V.N. Polymer Blends; *Khimiya*: Moscow, 1980 (in Russ).

[2] Krause S. In Polymer Blends, Paul D.R. and Newman S., Eds.; *Mir*: Moscow, 1981, Chap. 2 (Russ. Ed.).

[3] Wang R.; Russel R. *J. Interface Sci.* 2003, 11, 237.

[4] Kuleznev V.N. *Vysokomolek. Soedin.* 1993, A35, 1391.

[5] Kuleznev V.N.; Igosheva K.M. *Kolloid. Zhurn.* 1962, 24, 306.

[6] Krokhina L.S.; Kuleznev V.N.; Bukanova E.F. *Vysokomolek. Soedin.* 1974, A16, 1576.

[7] Assonova T.A.; Trapeznikov A.A. *Kolloid. Zhurn.* 1964, 26, 640.

[8] Kuleznev V.N.; Krokhina L.S. *Uspekhi Khimii* 1973, 42, 1278.

[9] Suthar V.; Pratar A.; Raval H. *Bull. Matter.* Sci. 2000, 23, 215.

[10] Horst R.; Wolf B.A. *J. Chem. Phys.* 1995, 103, 3782.

[11] Horst R.; Wolf B.A. *Macromol. Symp.* 1996, 112, 39.

[12] Kuhn R.; Bugdahl V.; Cantow H.-J. Angew. *Makromol. Chem.* 1971, 18, 115.

[13] Handbook of Chemistry. *Khimiya*: Leningrad 1966, Vol. 1 (in Russ).

[14] Scheffe H. J. Roy. *Statist. Soc.* 1958, B20, 344.

[15] Matsuo M.; Nozaki G.; Jio Y. *Polym. Eng. Sci.* 1969, 9, 97.

In: Preparation and Properties of Monomers, Polymers... ISBN 978-1-60021-557-5
Eds: A. Ballada and G. E. Zaikov, pp. 13-23

Chapter 2

REINFORCEMENT OF THE INTERFACE IN DRAWN POLYMER BLENDS PS/PA-12

Yu. P. Miroshnikov*, G. A. Chaves, S. V. Vlasov and V. I. Gerasimov#

M. V. Lomonosov Moscow State Academy of Fine Chemical Technology, 86 Vernadsky Ave., Moscow 117571, Russia

#M.V. Lomonosov Moscow State University. Leninskiye gory, Moscow 119899, Russia

ABSTRACT

The origination of the reinforcement effect in uniaxially drawn films on the basis of immiscible blends composed of crystalline matrix of polyamide-12 (PA) and glassy polystyrene (PS) dispersed phase has been investigated. Other conditions being equal, the effect becomes the more pronounced the lager the draw ratio and the finer the PS dispersion. Since the latter factor is itself a strong function of phase viscosity ratio (μ) the following two techniques of adjusting the optimal value of $\mu \approx 1$ have been adopted. According to the first one a selective plasticizer that is a good solvent for the PS inner phase and a non-solvent for the PA matrix phase was added to the more viscous PS. According to the second approach the value of μ was varied by the use of PS samples of different molecular weights. It was shown that significant reinforcement effect appears on drawing regardless of the method of adjusting optimal phase viscosity ratio. These data as well as the results obtained with the help of IR-dichroism and transmission electron microscopy evidence that the formation of much more ordered crystalline morphology of the PA matrix located at the PS/PA interfaces and the large interface area are mainly responsible for the discussed phenomena.

Keywords: reinforcement; polymer blends; drawn films; interface.

---

* Corresponding author: Yuri P. Miroshnikov. E-mail: miroshnikov@mitht.ru. Fax: +7 (495) 246-4741.Tel: +7 (495) 246-0958.

INTRODUCTION

In response to industrial demands many film polymeric materials have been designed. A large portion of these materials is prepared from immiscible polymer blends with different phase morphologies. Thorough tailoring of component properties, interfacial adhesion and phase morphology can lead to designing of novel heterogeneous materials with unique characteristics unattainable with homopolymers [1, 2].

As far as heterophase binary polymer blends are concerned, the third component is always present in such systems. It is an interphase layer formed at the surface of contact of alternative phases. Depending on interfacial characteristics the thickness of this layer can reach tens of nanometers, therefore its total volume in a blend can not be ignored [3]. Keeping in mind that characteristics of the interphase layer differ from these of bulk components, one may expect noticeable influence of its presence on final properties of compositions [4, 5].

In our previous paper [6] the reinforcement of crystalline matrices with glassy particles of polystyrene (PS) has been detected. The effect became more pronounced when PS phase was selectively plasticized to match its viscosity to that of the matrix, and the blend films were subjected to drawing.

The aim of this work is to get further insight into the mechanism of reinforcement of crystalline matrices by dispersions of glassy particles in drawn films.

EXPERIMENTAL

Materials

Commercial grades of polyamide-12 (PA) and block polymerized polystyrene PS0 as well as four samples of PS1-PS4 of different molecular weights (MW) synthesized in laboratory were used without any interference. Di-2-ethyl hexyl adipate (DEHA) has been chosen as a selective plasticizer for the PS inner phase. DEHA was shown to be a good solvent for PS and a non-solvent for PA.

Blend Preparation and Procedures

Two techniques of varying component viscosity ratio $\mu = \eta_{PS}/\eta_{PA}$ (where η_{PS} and η_{PA} are the apparent viscosities of the PS and PA phases) have been employed. The first one is performed by addition of different amounts of DEHA into the PS0 inner phase while the second one uses PS samples of different MW designated from PS0 to PS4 in Table 1.

All polymer samples were vacuum-dried at 80 ^{o}C overnight before use. Preliminary mixing PS0 with DEHA was carried out on a heated two-roll mill at 150 °C for 10 min. Blends of PA with plasticized (PSpl) or unplasticized PS0-PS4 polystyrenes were prepared in a co-rotating twin screw extruder ARP-20-150 at 190 °C. Extruded films of 80mm wide and 300μm thick with composition ratios PS/PA 5/95, 10/90, 20/80 and 70/30 were prepared. Uniaxial drawing of extruded films was carried out in a laboratory drawing machine at deformation rate of 1000 %/min and drawing temperature of 120 °C followed by thermal fixation. It was shown in preliminary experiments that these drawing conditions were optimal and ensured maximal mechanical properties of the films.

Table 1. Some rheological properties of PS0 containing DEHA, samples of PS of different molecular weight and their blends with PA. Shear rate is 10 c^{-1}

Property	DEHA content in PS0 [wt %]					PA	PS1	PS2	PS3	PS4
	0	5	10	20	30					
MFI (g/10 min)	2.8	6.1	9.8	22.1	98.8	14.8	5.8	13.5	25.2	113
η (MPa s)	1.20	0.51	0.32	0.14	0.04	0.20	0.63	0.23	0.11	0.033
$\mu=\eta_{PS}/\eta_{PA}$	6.00	2.55	1.60	0.70	0.20	-	3.15	1.15	0.55	0.17

The two-capillary technique and weight capillary rheometer were used to characterize rheological properties of polymers at 190 °C. Morphologies of films were analyzed by transmission electron microscopy (TEM) with the use of carbon replicas technique. Instron TM-1122 tensiometer was employed to measure tensile properties of blends in draw direction at room temperature and strain rate of 100 mm/min. Orientation parameters of PA phase were measured by birefringence and IR-dichroism techniques.

RESULTS AND DISCUSSION

On the basis of swelling behavior presented in Figure 1 a selective plasticizer for the blends has been selected. Obviously, DEHA dissolves PS even at room temperature, while the equilibrium swelling of PA in DEHA at 20°C and at 190°C were 1.7% and 2.3% respectively.

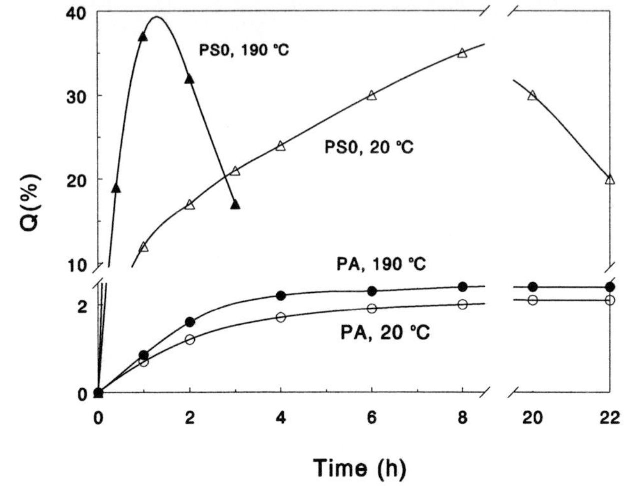

Figure 1. Kinetics of swelling of PA (1, 2) and PS0 (3, 4) in DEHA at 20 °C (1, 3) and 190 °C (2, 4).

Figure 2 illustrates the influence of DEHA concentration in the PS0 phase in blends 5/95 PSpl/PA (a) and the content of PSpl-phase (concentration of DEHA in PS0 is 20 wt.%) (b) on yield stress (σ_y), tensile strength (σ_b), elastic modulus (E), birefringence (Δn) and elongation at break (ε) for extruded films. It can be observed that molecular orientation and tensile properties reach slight maxima at 20 wt.% of DEHA in PS0 and at 5 wt.% of PSpl phase in blends with PA. The reinforcement effect grows noticeably when the blends are subjected to drawing. At draw ratio K=4 the tensile strength and elastic modulus of blends with the optimal compositions increase from 59 and 1300 MPa to 268 and 5400 MPa respectively (Figure 3). Moreover, the presence of PSpl dispersion enlarges the drawability from K=4 for the virgin PA matrix to K=5-6 for the blends.

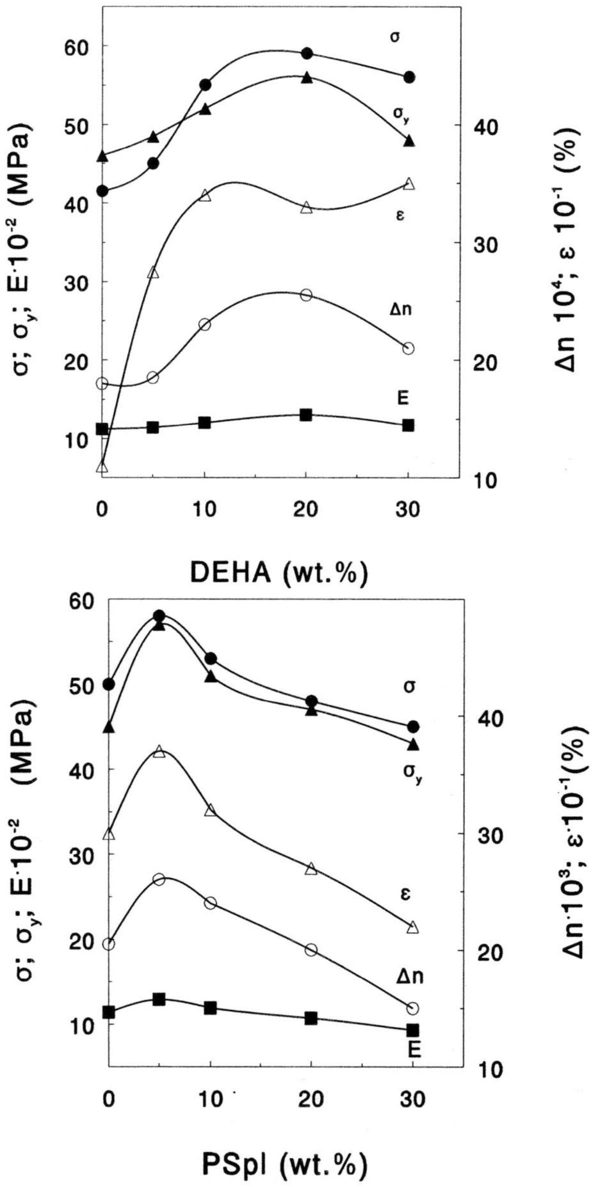

Figure 2. Dependences of tensile strength, σ, yield stress, σ_y, elastic modulus, E, elongation at break, ε, and birefringence, Δn, on DEHA concentration in PS0 (PSpl/PA 5/95) (a) and on the content of PSpl (20 wt% of DEHA) in blends with PA(b). Extruded films.

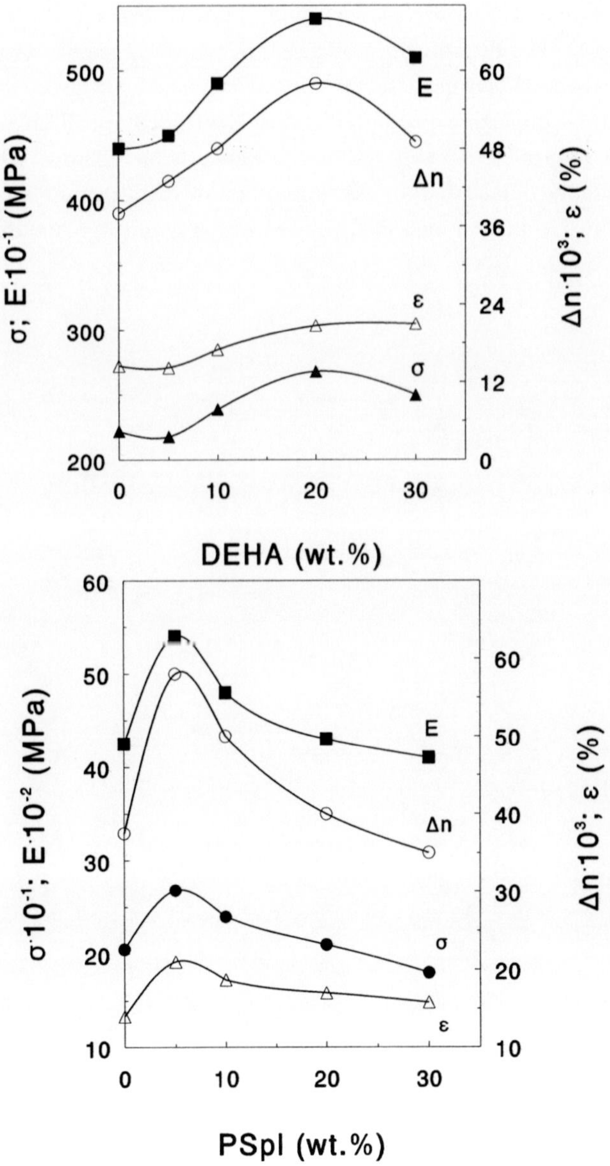

Figure 3. Dependences of σ, E, Δn and ε on DEHA concentration in PS0 (PSpl/PA 5/95) (a) and on PSpl (20 wt. % of DEHA) content in its blends with PA (b). Uniaxially drawn films, K=4.

The role of DEHA in plasticized PSpl/PA blends can be understood from the data in Table 1. It is clear that the increase of DEHA content makes PS0 less viscous so that at 20 wt.% of DEHA the viscosity of the PSpl phase becomes close to that of PA (μ=0.70). This value of phase viscosity ratio, as it was shown elsewhere [4], favors the formation of fine dispersion of the PSpl domains in the matrix of PA upon melt blending. This morphology is further optimized by the proper blend composition PSpl/PA 5/95 at which the smallest particles of the PSpl inner phase are formed. Thus, the results presented seem to indicate the dependence of the reinforcement effect on the scale of dispersion of the blends.

It was interesting to clarify the role of DEHA in the reinforcement phenomenon. With this aim the second series of oriented blends was prepared under similar conditions in which phase viscosity ratio was varied by changing MW of the PS phase in the absence of a plasticizer (Table 1). The dependences of σ, E, and Δn on phase viscosity ratio go through the maxima at μ~l when PS2 is used as the dispersed phase (not presented here). As Figure 4 shows the measured properties reach maximum values at composition ratio PS2/PA 5/95.

Analysis of Figures 2-4 leads to the following conclusions. The properties of both plasticized and unplasticized blends are optimized at composition ratio PS/PA 5/95 and at phase viscosity ratio μ~l when maximal scale of dispersion of the blends is attained.

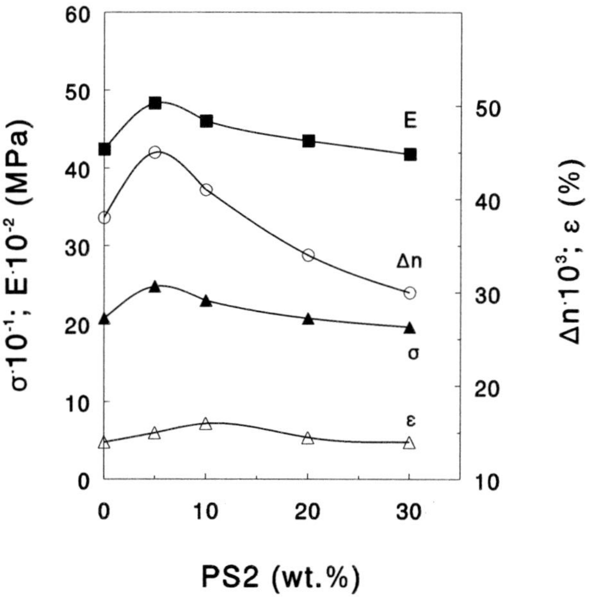

Figure 4. Dependences of σ, E, Δn and ε of unplasticized PS2/PA blends on PS2 content. Uniaxially drawn films, K=4.

Figure 5 representing the dependence of the average domain diameter of the PS inner phase (solid lines) and the specific interface area, S (dashed lines) for both types of extruded blends on phase viscosity ratio, μ, proves this conclusion. Curves 1 and 2 belongs to the blends obtained with plasticized and unplasticized polystyrenes respectively. Regardless of the manner of changing μ, the domain sizes go through the minima and S goes through the maxima when the viscosities of the phases are properly matched. These data strongly support the idea that the total interfacial area of blends plays an important role in the reinforcement phenomena discussed here.

On the other hand, the level of properties of drawn films attained with unplasticized PS of different MW is lower compared to the plasticized ones. Therefore, one can assume that DEHA not only helps to tailor the viscosities of the phases but affects properties of the matrix as well. To verify this assumption a series of blends of virgin PA with varying contents of DEHA were prepared and studied under conditions similar to those in the above experiments.

The addition of large amounts of DEHA into PA generally produces negative effects on tensile properties of the extruded films. In contrast, as Figure 6 illustrates, some reinforcement (from 205 to 230 MPa) of PA is attained at about 2.0 wt. % of DEHA in the drawn films.

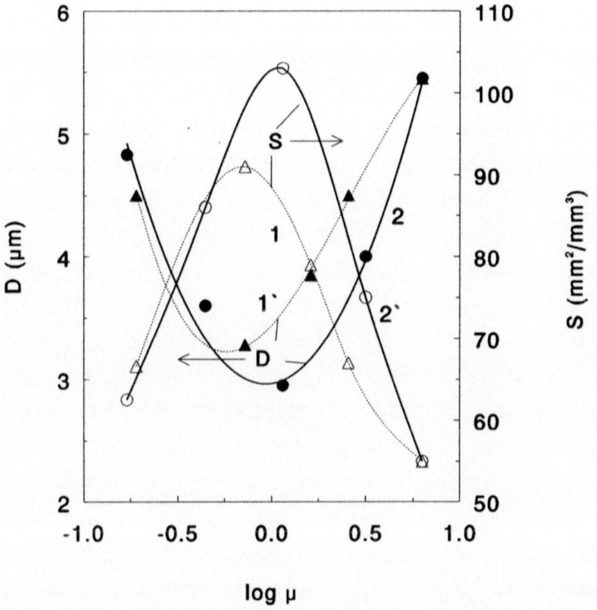

Figure 5. Average diameter (D) of PS domains (solid lines) and specific interface area (S) (dashed lines) versus phase viscosity ratio (μ) of blends. The value of μ was varied by changing DEHA content in PS0 (curves 1, 1') or by variation of molecular weight of PS samples (curves 2, 2').

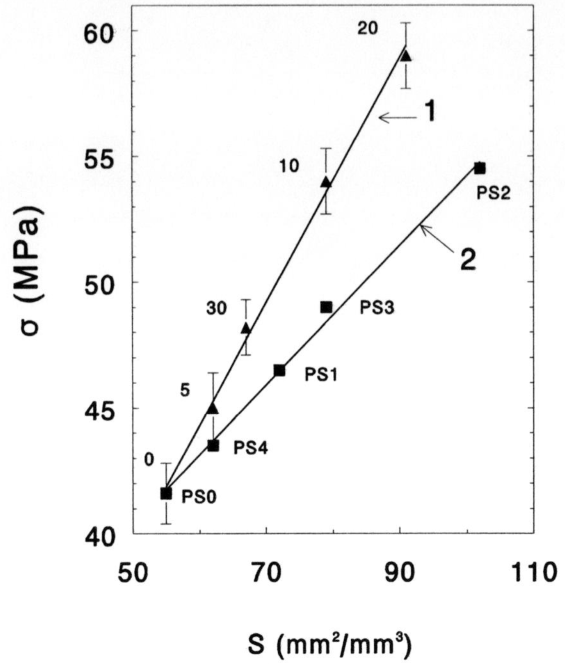

Figure 6. Influence of the content of DEHA on tensile strength, σ_b, elongation at break, ε, and birefringence, Δn, of the drawn PA matrix, K=4.

This is assumed to be a well known effect of the reinforcement action produced by small portions of plasticizer of poor thermodynamic quality which (at optimal dozes) favor the processes of orientation and crystallization of polymers.

The data presented above can be summarized in the following way. Generally, incompatible glassy domains in a crystalline matrix act like mechanical defects reducing the mechanical properties of blends. The reinforcement appears only in the drawn films (remember that the extruded films are slightly oriented) provided a small content of the inner phase and its smallest domain sizes are in place. The second condition is fulfilled when the viscosities of phases are matched properly upon blending. The best properties are achieved when selective plasticizer for the inner phase is used to equalize the viscosities of polymers. In this case the plasticizer itself additionally reinforces the matrix.

But it seems that the major role in the reinforcement of the drawn films belongs to the interfacial area. One can speculate that in oriented films the PS domains continue to act like defects but the positive effect of the reinforcement of the matrix overrides this negative contribution and leads to overall positive response. Figure 7 shows a practically linear growth of tensile strength of the blends upon increasing specific interface area. These data are obtained for the extruded films with low level of molecular orientation of the PA matrix. Unfortunately, the PS fiber-like domains in drawn films go too long to ensure accurate measurements of their volumetric sizes on microscopic images.

Figures 2-4 show that the birefringence, Δn, varies symbatically with mechanical responses. It means that the maximum level of mechanical properties is owed to the highest degree of molecular orientation of the matrix.

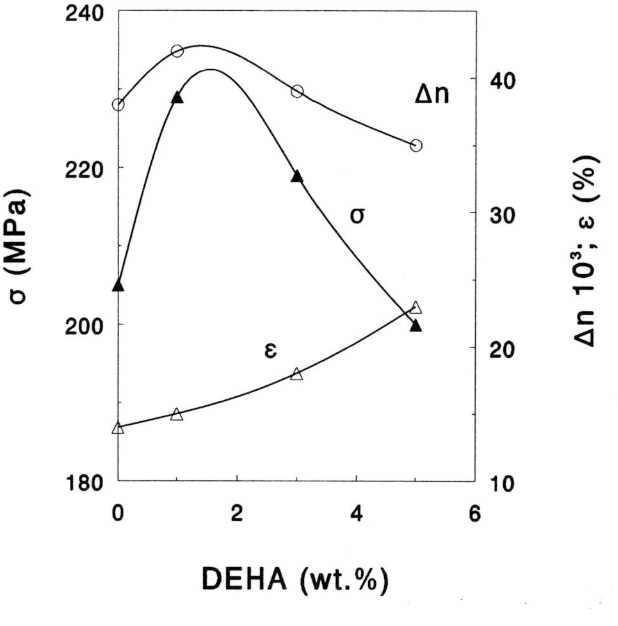

Figure 7. Tensile strength versus specific interface area for extruded 5/95 PS/PA blends containing PS0 loaded with different amounts of DEHA (curve 1) or PS samples of different molecular weight (curve 2). Concentration of DEHA in wt.% and PS samples are designated next to corresponding experimental points.

To get more information about this matter a method of IR-dichroism has been employed. Thin (ca. 80 μm) films of virgin PA and different blends containing plasticized PS were analyzed on the computerized IR-spectrometer UR-20 at wavenumbers 425 and 537 cm^{-1}. The data listed in Table 2 illustrate the growth of the dichroic ratio (R), average segment orientation angle (Φ) and Hermans function (F_y) with the draw ratio (K) of films and prove the results obtained with the birefringence technique.

Table 2. Orientation parameters for the PSpl/PA blends (numerator) and virgin PA (denominator) at different draw ratios: dichroic ratio (R); average segment orientation angle (Φ) and Hermans function (F_y)

K	R	Φ	F_v
1	2.02/1.56	45.8/49.0	0.25/0.14
2	4.17/3.46	34.4/36.9	0.52/0.46
3	8.99/7.49	24.8/28.6	0.73/0.68
4	16.48/11.79	18.7/21.9	0.85/0.79
5	22.82/ -	15.47 -	0.907 -

It should be emphasized here that the level of orientation of the PA matrix in the blends is 10-20% higher than that in virgin PA at the same draw ratios. It can be due to the presence of dissimilar PS particles increasing the local degree of orientation of the matrix. The results obtained support the orientational origin of the reinforcement effect which develops in the crystalline matrix. But the key point of the subject is that this effect is enhanced substantially when highly developed interface is formed.

To further clarify the role of the interface in the reinforcement mechanism, TEM studies of drawn films have been undertaken using the carbon replicas technique. The films were delaminated along the draw direction followed by the vapor carbon deposition on both fracture surfaces. The carbon replicas obtained were contrasted by aslant deposition of Al and analyzed by TEM at magnifications of up to 100.000. The micrographs shown in Figure 8 illustrate two-phase morphology of films represented by structureless amorphous highly elongated threadlike domains of plasticized PS embedded into the crystalline matrix of PA.

The principal feature of the employed technique of sample preparation is that owing to weak interfacial adhesion some part of the PSpl threads are eliminated from their places upon film splitting leaving naked interfaces (shown by arrows). Bulk matrix material is represented by the lamellae oriented normally to the draw direction. It can be seen that interlamellar distances on the interface are much smaller than those in the bulky PA. In other words, in contrast to the bulk portions of PA the crystalline structure on the interface looks like annealed and better organized.

It can be assumed from these data that chain segments of PA on the interface are more mobile compared to those in the bulk material. They form a more ordered and strong crystalline morphology which, keeping in mind a large interfacial area, makes a substantial contribution to mechanical response of the blends. This effect grows significantly upon drawing because both molecular orientation and interfacial area are progressively increased with draw ratio.

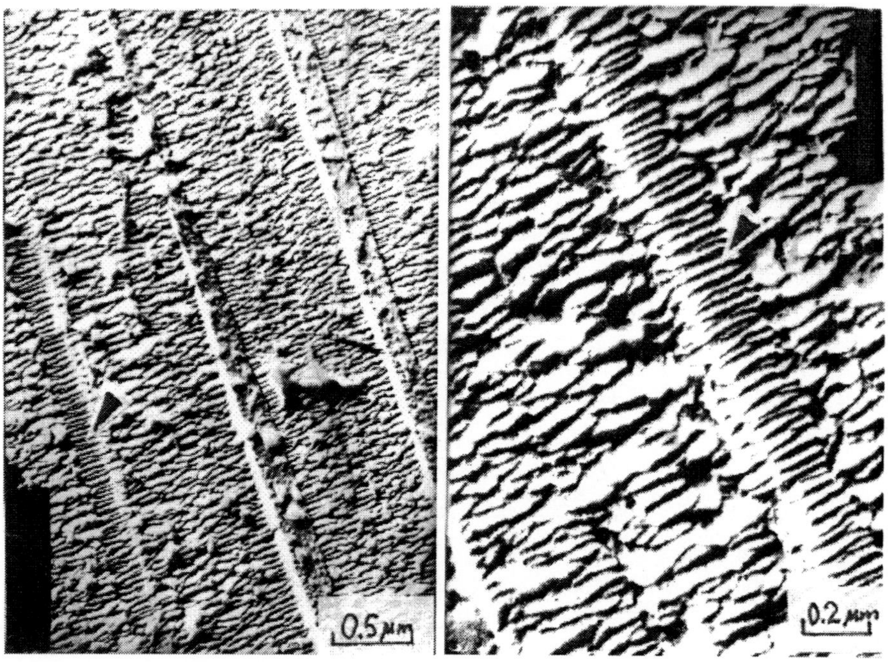

a) b)

Figure 8. TEM micrographs of carbon replicas made from fracture surfaces of 5/95 PSpl/PA drawn films, K=4. DEHA concentration in PS0 is 5 wt.%: (a) - general picture illustrating the morphology of a blend with thread-like PSpl domains embedded into a crystalline matrix of PA. One of the threads is absent baring the interface (pointed by arrow); (b) – a closer view of the previous image showing the PSpl/PA interface (pointed by arrow). Notice the difference between crystalline textures on the interface and in the bulk PA.

Because the reinforcement effect pertains to both selectively plasticized and unplasticized PS/PA blends it is clear that the plasticizer itself does not produce this effect but enhances it according to the mechanism mentioned above. Considering very weak interfacial adhesion in studied blends it can be assumed that the formation of highly developed interface and proper orientation are the main conditions which can produce similar reinforcement phenomena in other crystalline matrices irrespectively of the nature of the inner phase.

CONCLUSIONS

The mechanism of reinforcement of crystalline matrix of PA-12 by dispersions of glassy particles of PS in drawn films has been proposed and discussed. The highest reinforcement effect is observed in drawn films provided composition ratio and phase viscosities (i.e. $\mu \sim 0.7$) are matched properly to ensure the highest interfacial area possible. The presence of a plasticizer enhances the effect.

The main features of the mechanism proposed are as follows. Chain segments at the interface seem to have higher mobility compared to that in the bulk matrix, especially in presence of a plasticizer. This favors the formation of a more ordered and strong crystalline regions in PA at the interface. The effect is enhanced noticeably by drawing and increasing the interface area.

ACKNOWLEDGEMENTS

The authors express their gratitude to Mr. V.V. Matveev for his help in TEM analysis.

REFERENCIES

[1] Markov A.V. *Plast. Massy* 2005, 1, 33.

[2] Vlasov S.V.; Sherbakova E.A.; Markov A.V.; Kuleznev V.N. Int. *J. Polym. Mater.* 2004, 53, 685.

[3] *Polymer blends* vol. 1. D.R. Paul and S. Newman Eds., Academic Press: New York, 1978.

[4] Kuleznev V.N. Polymer blends. *Khimiya*: Moscow, 1980 (in Russ).

[5] Sakellar S.L.; McHugh A.I. *Rheol. Acta* 1987, 26, 64.

[6] Chaves G.A.; Miroshnikov Yu.P.; Kuleznev V.N.; Vlasov S.V.; Rukutuariveli G. *Plast. Massy* 1991, 1, 12.

In: Preparation and Properties of Monomers, Polymers... ISBN 978-1-60021-557-5
Eds: A. Ballada and G. E. Zaikov, pp. 25-34

Chapter 3

QUANTUM CHEMICAL CALCULATION LINEAR OLEFINS AND NOT CONJUGATE DIOLEFINS

V. A. Babkin*, A. N. Meshkov, S. M. Lomakin, Ju. A. Sangalov and G. E. Zaikov
Institute of Biochemical Physics, Russian Academy of Sciences,
4, Kosygin st., 119991 Moscow, Russia

ABSTRACT

Quantum-chemical method CNDO/2 in parametrization Santri-Popla-Segala executes calculation of some linear olefins and not conjugate diolefins C_2H_4, C_3H_7, C_4H_8, C_5H_{10}, C_6H_8, C_7H_{14}, C_8H_{16}, $C_{10}H_{20}$. The optimized geometrical and electronic structure of these connections is received. Their acid strength is appreciated. All is established, that linear olefins and diolefins, irrespective of a degree branching have identical and rather low acid strength pKa = + 41,79.

Keywords: Quantum-chemical calculation, method CNDO/2, a rating of acid strength, linear olefins and not conjugate diolefins.

AIMS AND BACKGROUND

Process of reception of traditional polymers (polyethylene, polypropylene, polyisobutylene and others which are widely used in the industry of building materials - PSM) from linear and not conjugate monomers - rather complex catalytic process cationic polymerizations [1, 2]. Usually it includes 3 stages: initiation, growth and breakage of a material circuit. Realization of these reactions guarantees presence basically 4 components: the catalyst, promotor (socatalyst), solvent and a monomeasure - not including every possible

* VolgSABU, 21 Michurina st., 403343, Michailovka Volgograd region, Russia. E-mail: chembio@sky.chph.ras.ru

additives and inhibitors. Influence of acid strength of the initiator (promotor and the catalyst) on efficiency of process of polymerization at present time is well enough investigated and rather rigidly conjugate to selectivity [1, 2]. A question of influence of acid strength monomers on chemical reactions cationing polymerizations olefins, a rating of their acid strength till now remaining open.

In this connection, the purpose of the present work is the theoretical rating of acid strength various linear olefins and not conjugate diolefins through quantum-chemical calculation of some typical monomers: C_2H_4, C_3H_7, C_4H_8, C_5H_{10}, C_6H_8, C_7H_{14}, C_8H_{16}, $C_{10}H_{20}$.

METHODICAL PART

Quantum-chemical calculation linear monomers and not conjugate diolephins carried out classical halfempiric method CNDO/2 in parametrization Santri-Popla-Segala [3]. Acid strength researched olephins it was estimated under the formula $pKa = 43.7\text{-}190.3 \bullet q^{H+}{}_{max}$ (1) (where pKa - universal parameter of an acid, $q^{H+}{}_{max}$ - charge on atom of hydrogen [1]).

RESULTS OF CALCULATIONS AND THEIR DISCUSSIONS

Geometrical and electronic structure linear olefins and not conjugate diolefins, and also some quantum - chemical parameters and their acid strength are submitted in tables 1-9 and on Figures 1-8.

Table 1. Lengths of bonds and valent angles ethylene

Lengths of bond	R, A	Valent angles	Hailstones
C(1)-C(2)	1.34	H(3)-C(1)-C(2)	123
C(1)-H(3)	1.08	H(4)-C(1)-C(2)	123
C(1)-H(4)	1.08	H(6)-C(2)-C(1)	123
C(2)-H(5)	1.08	H(5)-C(2)-C(1)	123
C(2)-H(6)	1.08		

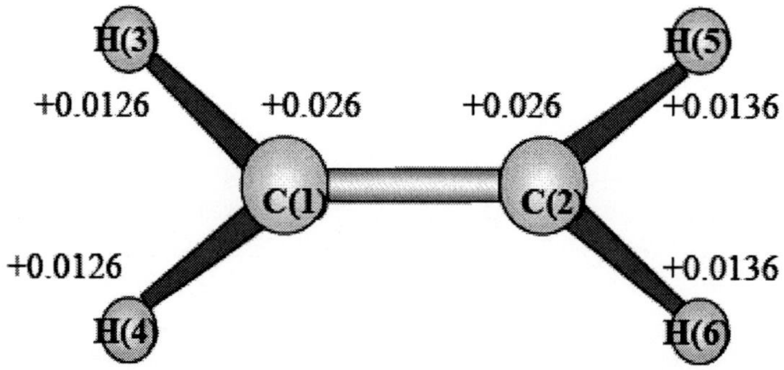

Figure 1. Geometrical and electronic structure ethylene. (E_0= -17.07 A.U., E_{CB} = -2.18 A.U. ,D = 0.00 dB).

Table 2. Lengths of bond and valent angles of propylene

Lengths of bond	R, A	Valent angles.	Hailstones
C(1)-C(2)	1.34	H(3)-C(1)-C(2)	117
C(2)-C(3)	1.46	H(4)-C(1)-C(2)	117
C(1)-H(4)	1.12	H(6)-C(2)-C(1)	116
C(1)-H(5)	1.12	C(1)-C(2)-C(3)	118
C(2)-H(6)	1.12	H(7)-C(3)-C(2)	113
C(3)-H(7)	1.12	H(8)-C(3)-C(2)	114
C(3)-H(8)	1.12	H(9)-C(3)-C(2)	113
C(3)-H(9)	1.12		

Table 3. Lengths of bonds and valent angles olefins C_4H_8 (2)

Lengths of bonds	R, A	Valent angles	Hailstones.
C(1)-C(2)	1.33	C(1)-C(2)-C(3)	127
C(2)-C(3)	1.45	C(2)-C(3)-C(4)	113
C(3)-C(4)	1.46	H(4)-C(1)-C(2)	112
C(1)-H(4)	1.00	H(5)-C(1)-C(2)	127
C(1)-H(5)	1.12	H(6)-C(1)-C(2)	119
C(2)-H(6)	1.12	H(7)-C(2)-C(1)	113
C(3)-H(7)	1.12	H(8)-C(3)-C(2)	111
C(3)-H(8)	1.12	H(9)-C(3)-C(2)	112
C(3)-H(9)	1.12	H(10)-C(4)-C(3)	112
C(4)-H(10)	1.12	H(11)-C(4)-C(3)	112
C(4)-H(11)	1.12	H(12)-C(4)-C(3)	112
C(4)-H(12)	1.12		

Table 4. Lengths of bonds and valent angles olefins C_4H_8 (3)

Lengths of bonds	R, A	Valent angles	Hailstones
C(1)-C(2)	1.32	C(3)-C(1)-C(2)	117
C(1)-C(3)	1.45	C(4)-C(2)-C(1)	123
C(2)-C(4)	1.45	H(5)-C(1)-C(2)	129
C(1)-H(5)	1.11	H(6)-C(1)-C(2)	121
C(2)-H(6)	1.12	H(7)-C(3)-C(1)	112
C(3)-H(7)	1.12	H(8)-C(3)-C(1)	111
C(3)-H(8)	1.12	H(9)-C(3)-C(1)	111
C(3)-H(9)	1.12	H(10)-C(4)-C(2)	113
C(4)-H(10)	1.12	H(11)-C(4)-C(2)	112
C(4)-H(11)	1.12	H(12)-C(4)-C(2)	112
C(4)-H(12)	1.12		

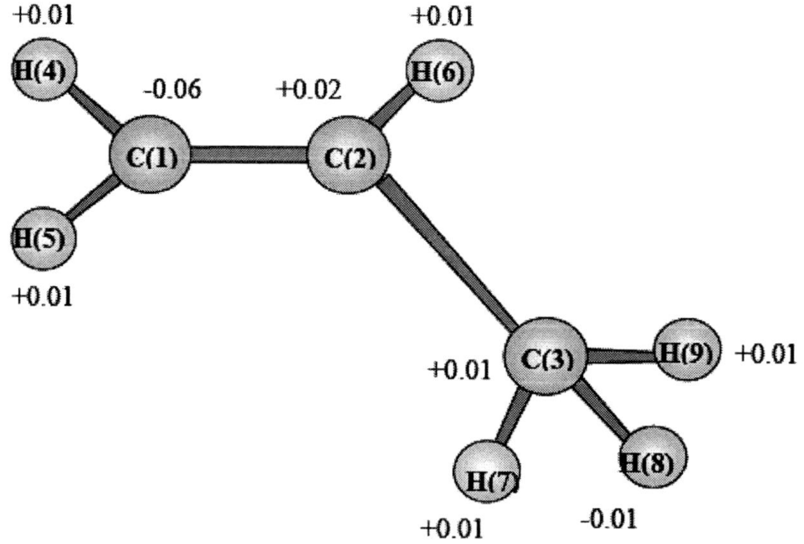

Figure 2. Geometrical and electronic structure propylene. (E_0 = -25.77 A.U., E_{CB} = -3.44 A.U., D = 0.26 dB).

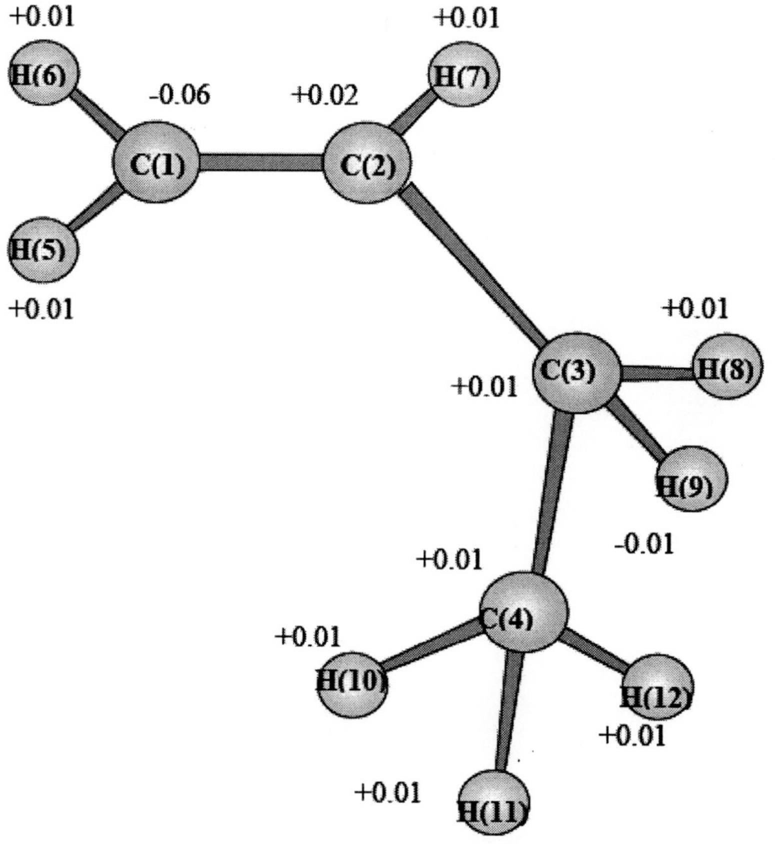

Figure 3. Geometrical and electronic structure olefins C_4H_8(2). (E_0 = -34.49 A.U., E_{CB} = -4.71 A.U., D = 0.39 dB).

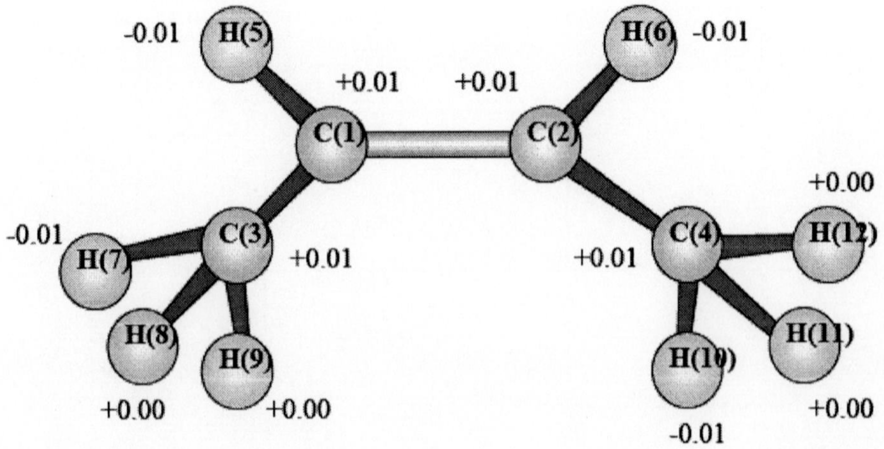

Figure 4. Geometrical and electronic structure olefins C_4H_8.(3). (E_0 = -34.4966 A.U., E_{CB} = -4.7268 A.U., D = 0.22 dB).

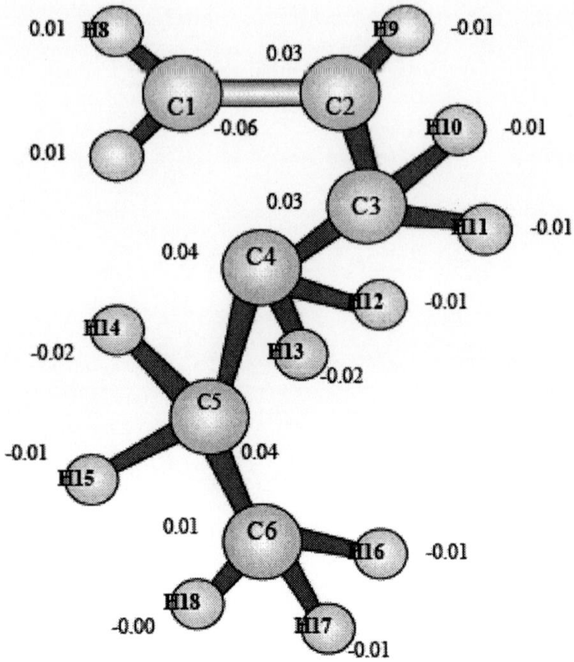

Figure 5. Geometrical and electronic structure olefins C_6H_{12}. (E_0 = -51.88 A.U., E_{CB} = -7.23 A.U., D = 0.36 dB).

Table 5. Lengths of bond and valent angles olefins C_6H_{12}

Lengths of bonds	R, A	Valent angles	Hailstones
C(1)-C(2)	1.31	C(3)-C(2)-C(1)	126
C(2)-C(3)	1.47	C(4)-C(3)-C(2)	113
C(3)-C(4)	1.48	C(5)-C(4)-C(3)	114
C(4)-C(5)	1.48	C(6)-C(5)-C(4)	117
C(5)-C(6)	1.47	H(7)-C(1)-C(2)	122
C(1)-H(7)	1.11	H(8)-C(1)-C(2)	120
C(1)-H(8)	1.12	H(9)-C(2)-C(1)	119
C(2)-H(9)	1.12	H(10)-C(3)-C(2)	112
C(3)-H(10)	1.12	H(11)-C(3)-C(2)	110
C(3)-H(11)	1.12	H(12)-C(4)-C(3)	111
C(4)-H(12)	1.12	H(13)-C(4)-C(3)	112
C(4)-H(13)	1.12	H(14)-C(5)-C(4)	111
C(5)-H(14)	1.12	H(15)-C(5)-C(4)	110
C(5)-H(15)	1.12	H(16)-C(6)-C(5)	109
C(6)-H(16)	1.12	H(17)-C(6)-C(5)	111
C(6)-H(17)	1.12	H(18)-C(6)-C(5)	112
C(6)-H(18)	1.12		

Table 6. Lengths of bonds and valent angles olefins C_7H_{14}

Lengths of bonds	R, A	Valent angles	Hailstones
C(1)-C(2)	1.36	C(3)-C(2)-C(1)	127
C(2)-C(3)	1.48	C(4)-C(3)-C(2)	113
C(3)-C(4)	1.49	C(5)-C(4)-C(3)	114
C(4)-C(5)	1.49	C(6)-C(5)-C(4)	118
C(5)-C(6)	1.49	C(7)-C(6)-C(5)	115
C(6)-C(7)	1.48	H(8)-C(1)-C(2)	122
C(1)-H(8)	1.10	H(9)-C(1)-C(2)	121
C(1)-H(9)	1.12	H(10)-C(2)-C(1)	119
C(2)-H(10)	1.12	H(11)-C(3)-C(2)	113
C(3)-H(11)	1.12	H(12)-C(3)-C(2)	111
C(3)-H(12)	1.12	H(13)-C(4)-C(3)	112
C(4)-H(13)	1.12	H(14)-C(4)-C(3)	112
C(4)-H(14)	1.12	H(15)-C(5)-C(4)	112
C(5)-H(15)	1.12	H(16)-C(5)-C(4)	111
C(5)-H(16)	1.12	H(17)-C(6)-C(5)	109
C(6)-H(17)	1.12	H(18)-C(6)-C(5)	112
C(6)-H(18)	1.12	H(19)-C(7)-C(6)	112
C(7)-H(19)	1.12	H(20)-C(7)-C(6)	112
C(7)-H(20)	1.12	H(21)-C(7)-C(6)	112
C(7)-H(21)	1.12		

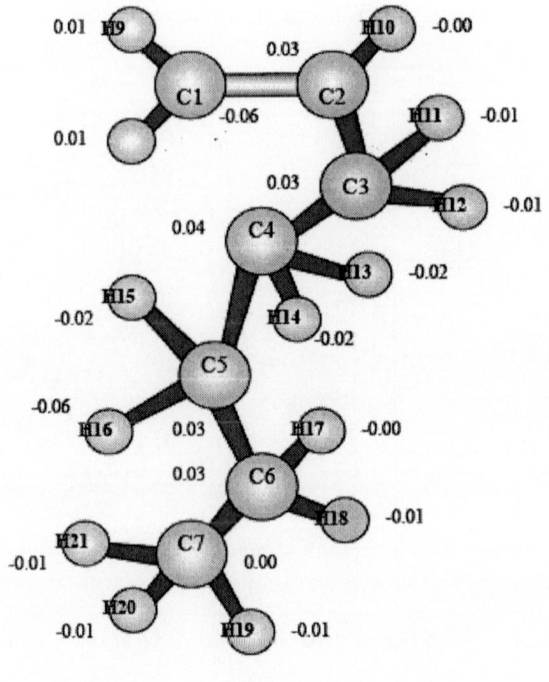

Figure 6. Geometrical and electronic structure olefins C_7H_{14}. (E_0 = -60.58 A.U., E_{CB} = -8.48 A.U., D = 0.36 dB.).

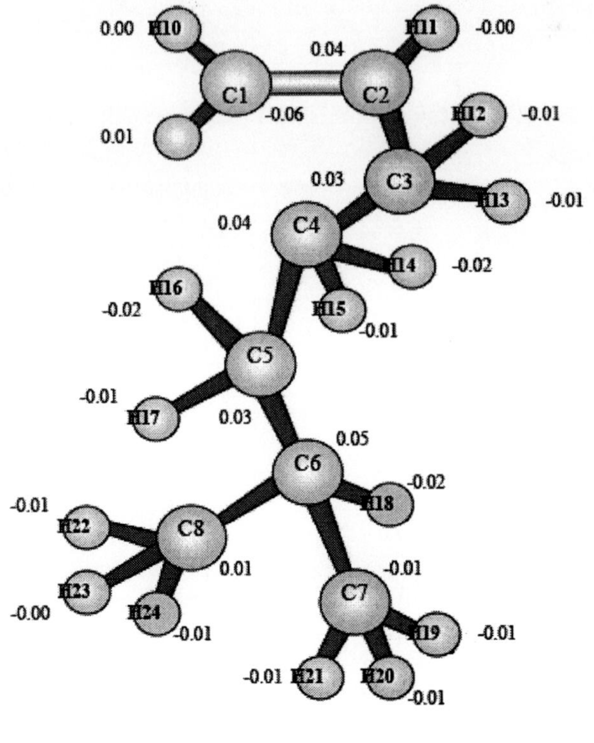

Figure 7. Geometrical and electronic structure olefins C_8H_{16}. (E_0 = -69.2611 A.U., E_{CB} = -9.72 A.U., D = 0.38 dB).

Table 7. Lengths of bonds and valent angles olefins C_8H_{16}

Lengths of bonds	R, A	Valent angles	Hailstones
C(1)-C(2)	1.32	C(3)-C(2)-C(1)	127
C(2)-C(3)	1.45	C(4)-C(3)-C(2)	112
C(3)-C(4)	1.49	C(5)-C(4)-C(3)	117
C(4)-C(5)	1.49	C(6)-C(5)-C(4)	121
C(5)-C(6)	1.51	C(7)-C(6)-C(5)	117
C(6)-C(7)	1.48	C(8)-C(6)-C(5)	110
C(6)-C(8)	1.49	H(9)-C(1)-C(2)	122
C(1)-H(9)	1.10	H(10)-C(1)-C(2)	118
C(1)-H(10)	1.12	H(11)-C(2)-C(1)	114
C(2)-H(11)	1.12	H(12)-C(3)-C(2)	110
C(3)-H(12)	1.12	H(13)-C(3)-C(2)	111
C(3)-H(13)	1.12	H(14)-C(4)-C(3)	111
C(4)-H(14)	1.12	H(15)-C(4)-C(3)	110
C(4)-H(15)	1.12	H(16)-C(5)-C(4)	111
C(5)-H(16)	1.12	H(17)-C(5)-C(4)	111
C(5)-H(17)	1.12	H(18)-C(6)-C(5)	112
C(6)-H(18)	1.12	H(19)-C(7)-C(5)	115
C(7)-H(19)	1.12	H(20)-C(7)-C(6)	114
C(7)-H(20)	1.12	H(21)-C(7)-C(6)	113
C(7)-H(21)	1.12	H(22)-C(8)-C(6)	111
C(8)-H(22)	1.12	H(23)-C(8)-C(6)	111
C(8)-H(23)	1.12	H(24)-C(8)-C(6)	111
C(8)-H(24)	1.12		

Table 8. Lengths of bonds and valent angles olefins $C_{10}H_{20}$

Lengths of bonds.	R, A	Valent angles	Hailstones
C(1)-C(2)	1.36	C(3)-C(2)-C(1)	127
C(2)-C(3)	1.48	C(4)-C(3)-C(2)	113
C(3)-C(4)	1.49	C(5)-C(4)-C(3)	114
C(4)-C(5)	1.49	C(6)-C(5)-C(4)	118
C(5)-C(6)	1.49	C(7)-C(6)-C(5)	115
C(6)-C(7)	1.48	H(8)-C(1)-C(2)	122
C(1)-H(8)	1.10	H(9)-C(1)-C(2)	121
C(1)-H(9)	1.12	H(10)-C(2)-C(1)	119
C(2)-H(10)	1.12	H(11)-C(3)-C(2)	113
C(3)-H(11)	1.12	H(12)-C(3)-C(2)	111
C(3)-H(12)	1.12	H(13)-C(4)-C(3)	112
C(4)-H(13)	1.12	H(14)-C(4)-C(3)	112
C(4)-H(14)	1.12	H(15)-C(5)-C(4)	112
C(5)-H(15)	1.12	H(16)-C(5)-C(4)	111
C(5)-H(16)	1.12	H(17)-C(6)-C(5)	109
C(6)-H(17)	1.12	H(18)-C(6)-C(5)	112
C(6)-H(18)	1.12	H(19)-C(7)-C(6)	112
C(7)-H(19)	1.12	H(20)-C(7)-C(6)	112
C(7)-H(20)	1.12	H(21)-C(7)-C(6)	112
C(7)-H(21)	1.12		

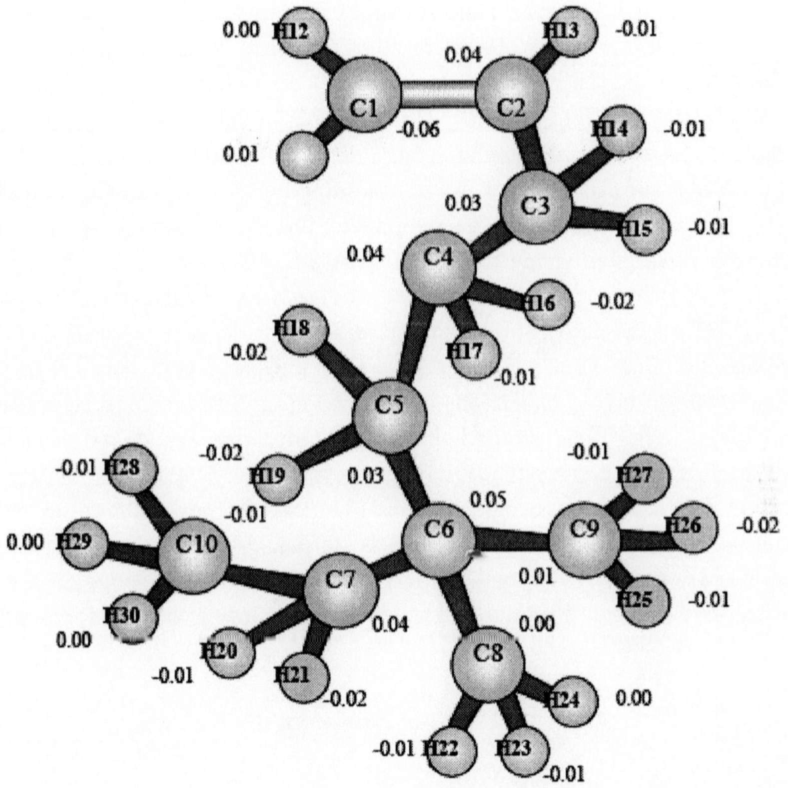

Figure 8. Geometrical and electronic structure olefins $C_{10}H_{20}$. (E_0 = -86.63 A.U., E_{CB} = -12.20 A.U., D = 0.35 dB).

Table 9. Quantum-chemical parameters and acid strength linear olefins and not conjugate diolefins (see figures 1-8, tables 1-8)

№	Olefins	Eo (A.U.)	EcB(A.U.)	D (dB)	q(c1)	q(c2)	pKa	q^{H+}_{max}
1	C2H4	-17.07	-2.18	0	0.02	0.02	41.79	0.01
2	C3H7	-25.77	-3.44	0.26	0.06	0.02	41.79	0.01
3	C4H8 (2)	----	----	----	-0.06	0.02	41.79	0.01
4	C4H8 (3)	-34.5	-4.73	0.22	0.01	0.01	41.79	0.01
5	C6H8	-51.88	-7.23	0.36	-0.06	0.03	41.79	0.01
6	C7H14	-60.55	-8.48	0.36	-0.06	0.03	41.79	0.01
7	C8H16	-69.26	-9.72	0.38	-0.06	0.04	41.79	0.01
8	C10H20	-86.63	-12.2	0.35	-0.06	0.04	41.79	0.01

From these tables it is visible, that at all investigated monomers the maximal charge on atom of hydrogen q^{H+} identical and is equal +0.01. From here, it is easy to define under the formula (1) values pKa. It is obvious, that they also are equal among themselves, and values pKa = + 41.79. In this connection, if artificial, prior to the beginning of initiation of polymerization to not change acid strength the monomeasure introduction of various additives as H-acids process of polymerization will not depend neither on acid strength a monomeasure and nor from his(its) nature and from a degree branching.

Also from table 9 it is visible, that propensity olefins to polymerization on cations to the mechanism according to rule Morcovnik-the classical mechanism of process [1], the monomeasure, and size of a negative charge on a - carbon atom – q^{H+} a substratum will be defined not brenstedovky acid strength. This tendency is rather precisely traced in a number olefins ethylene, propylene, isobutylene: qc a (C_2H_4) = + 0,02; qc a (C_3H_7) = -0.06, qc a (C_4H_8) = -0.07, that corresponds to a fact of common knowledge: ethylene in usual conditions with the big work polymerization on cations to the mechanism, and isobutylene on the contrary has the maximal predisposition in this sense [2, 4].

Thus, for the first time us was executed quantum-chemical calculation halfempiric by method CNDO/2 in parametrization Santri-Popla-Segala of some typical linear olefins and not conjugate diolefins: C_2H_4, C_3H_7, C_4H_8, C_7H_{14}, C_8H_{16}, $C_{10}H_{20}$. The geometrical and electronic structure of these connections is received by optimized standard gradient a method on all parameters. Their acid strength is theoretically appreciated. All is established, that researched olefins and conjugate diolefins, irrespective of a degree of branching have identical and rather low acid strength pKa = + 41,79. It is shown, that propensity olefins to polymerization on cation to the mechanism is defined by size of a negative charge on a carbon atom of a substratum. And the tendency is traced: the more value qc a, the propensity to polymerization olefins on cation to the mechanism according to rule Morcovnik - is more.

REFERENCES

[1] Babkin V.A. *Structure of the active centers*. The mechanism of initiation and growth of a circuit at cation polymerizations olefins at the presence of complex catalysts. 1999, Moscow, 214 pp. The thesis for degree of Doctor science. Institute of Chemical Physics Russian Academy of Sciences.

[2] Sangalov J.A., Minsker K.S. *Polymers and copolymers of isobutylene*. 2001, Agidol Publishers, Ufa, 381pp.

[3] Sigal J. *Halfempiric methods of calculation of electronic structure*. 1980, Mir Publishers, Moscow, 327 pp.

[4] Kennedy J. *Cationic polymerization olefins*. 1978, Mir Publishers, Moscow, 431 pp.

In: Preparation and Properties of Monomers, Polymers... ISBN 978-1-60021-557-5
Eds: A. Ballada and G. E. Zaikov, pp. 35-37

Chapter 4

TECHNOLOGY COMPUTERS SEARCH OF NEW MORE EFFECTIVE CATALYSTS CATIONIC POLYMERIZATION OLEFINS

V. A. Babkin*, G. E. Zaikov and K. S. Minsker
Institute of Biochemical Physics, Russian Academy of Sciences,
4, Kosygin st., 117997 Moscow, Russia

ABSTRACT

For the first time the technology computers search for new more effective catalysts cationic polymerization olefins through quantum chemical calculation of modeling connections.

Keywords: new computers technology, catalysts, cation polymerization olefins, quantum chemical calculation, method CNDO/2.

AIMS AND BACKGROUND

For reception of polymers ethylene, propylene, izobutylene and others izoolefins, it is wide used in the industry of building materials (ISM), complex catalysts rather are frequently applied on the basis of Lewis's acids and Brensted's such as $RnAlCl_3 \cdot H_2O \cdot$(HCl, ROH), $RnBF_3 \cdot H_2O \cdot$(HF, ROH), $RnMgCl_2 \cdot H_2O \cdot$(HCL, ROH) and also their combinations and complex salts [1].

Synthesis of polymers - complex toilful, catalytic process. Selection of new more active and non-polluting initiators cationic polymerizations olefins- it is actual the problem representing doubtless scientific theoretical and practical interest.

* VolgSABU, 21 Michurina st., 403343, Michailovka Volgograd region, Russia; E-mail: chembio@sky.chph.ras.ru
Bashkirian State University, 32 Frunze st., 450074 Ufa, Russia

In this connection, the purpose of the present work is development and the description of algorithm of technology of computer search of new more effective initiators cationic polymerizations olefins through quantum-chemical calculation of models of tested catalysts.

METHODICAL PART AND RESULTS

Offered Algorithm of Computer Search

(1) Preparation of initial given tested catalysts (1,2,3, ...n) a bat-file for quantum-chemical calculations as a package of models.
(2) Quantum-chemical calculation of models catalytic systems, as H-acids anyone, for example, halfempiric method CNDO/2.
(3) Choice from quantum chemical calculation of the maximal charge on atom of hydrogen - q_{H+}^{max} of the catalyst.
(4) Rating of acid force catalytic systems, as ì-acids under the formula: pKa = 43.7-190.3•q^{H+}_{max} (where, pK_a – universal index of acidity [1]).
(5) Theoretical rating of efficiency of the tested catalyst (pKa ≤ =+1.0-catalyst active, but not selective; 1,0 ≤ pKa ≤ + 14.4-catalyst active and selective; pKa ≥ + 14.4 - the complex is not or is the bad catalyst cationic polymerizations olefins) and a choice of the best.
(6) Experimental check of complex catalysts on activity, selectivity and ecological cleanliness.

The block -scheme of algorithm of technology of search of new more effective catalysts cationic polymerizations olefins, used in ISM.

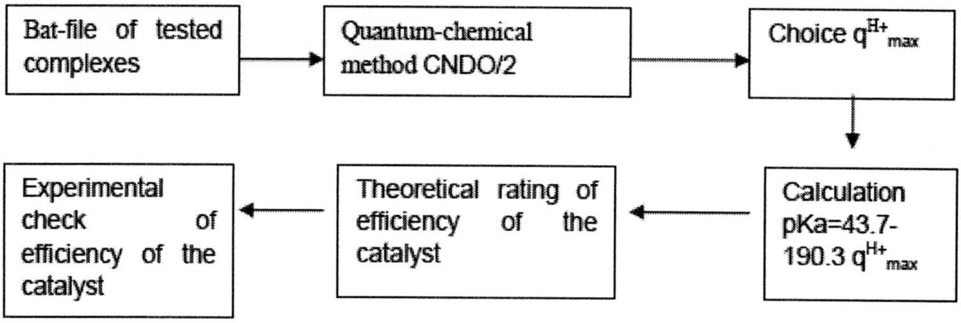

CONCLUSIONS

Such the images, the offered algorithm of technology of computer search of new more effective catalysts cationic polymerization olefins from huge variety of every possible complex systems and their combinations to choose enough quickly, the most effective for experimental check, that in turn will open an opportunity for synthesis catalytic systems which can promote reception of polymers with preset properties.

REFERENCES

[1] Babkin V.A. *Structure of the active centers. The mechanism of initiation and growth of a circuit at cation polymerizations olefins at the presence of complex catalysts*. 1999, Moscow, 214 pp. The thesis for degree of Doctor science. Institute of Chemical Physics Russian Academy of Sciences.

In: Preparation and Properties of Monomers, Polymers… ISBN 978-1-60021-557-5
Eds: A. Ballada and G. E. Zaikov, pp. 39-56

Chapter 5

Quantum Chemical Calculation and an Estimation of Acid Force Linear and Ramified Connected Diens

V. A. Babkin*, A. V. Kudryashov, Ju. A. Sangalov and G. E. Zaikov
Institute of Biochemical Physics, Russian Academy of Sciences,
4, Kosygin st., 119991 Moscow, Russia

Abstract

Quantum chemical method CNDO/2 in parameterization Santry-Popla-Seagala executes calculation of some linear and ramified diens: butadien-1,3; piperilen; isoprene; cis, trans - gexadien-2,4; cis, cis - gexadien -2,4; trans, trance - gexadien-2,4; 1,3-dimethylbutadien-1,3; 2,3-dimethylbutadien-1,3; cis -3-methylpentadien-1,3; trans - 3-methylpentadien-1,3; 4-methylpentadien-1,3; 1,1,3-trimethylbutadien-1,3; 1,1,4,4-tetramethylbutadien-1,3; 2-ethylbutadien-1,3; 2-tret-buthylbutadien-1,3. The optimized geometrical and electronic structure of these connections is received. It is established, that linear and ramified connected diens, irrespective of a degree of branching have rather low acid force pKa = + 36.1 — +43.7.

Full enough information about cationic polymerization of linear and ramified connected diens is submitted in rather serious critical review [1]. However, till now it is not investigated mechanisms of initiation, growth and breakage of a circuit of these monomers at an electronic level. These are not clear questions of influence of acid force of initiators (promoter and the catalyst), solvents and connected diens on efficiency of process of cationic polymerization, and in particular, on selectivity of reactions.

In this connection, the purpose of the present work is the theoretical estimation of acid force linear and ramified connected diens butadien-1,3; piperilen; isoprene; cis, trans -gexadien-2,4; cis, cis - gexadien -2,4; trans, trance - gexadien-2,4; 1,3-dimethylbutadien-1,3; 2,3-dimethylbutadien-1,3; cis -3-methylpentadien-1,3; trans - 3-methylpentadien-1,3; 4-methylpentadien-1,3; 1,1,3-trimethylbutadien-1,3; 1,1,4,4-tetramethylbutadien-1,3; 2-ethylbutadien-1,3; 2-tret-buthylbutadien-1,3 through quantum chemical calculation of these monomers, that can promote deeper understanding of thin mechanisms of cationic

* E-mail: sfi@reg.avtlg.ru; E-mail: chembio@sky.chph.ras.ru

polymerization of connected linear and ramified diens, and in a particular, at an electronic level.

Methodical Part

Quantum chemical calculation of linear and ramified connected diens is executed by classical half empiric method CNDO/2 in parameterization Santry-Popla-Seagala [2]. Acid force of researched diens was estimated under the known formula pKa = 43.7-190.3•q^{H+}_{max} (where pKa - universal parameter of acidity, the q^{H+}_{max}- as much as possible charge on atom of hydrogen [3]).

Results of Calculations and Their Discussions

Geometrical and electronic structure of linear and ramified connected diens, and also some quantum chemical parameters and their acid force are submitted in tables 1-16 and on figures 1-15.

Table 1. Lengths of bonds and valent corners butadien-1,3

Lengths of bonds	R,A	Valent corners	Degree
C(2)-C(1)	1.33	H(3) -C(1)-C(2)	120
C(1)-H(3)	1.12	H(4)-C(1)-C(2)	122
C(1)-H(4)	1.12	H(5)-C(2)-C(1)	119
C(2)-H(5)	1.12	H(7)-C(6)-C(2)	117
C(2)-C (6)	1.44	C(6)-C(2)-C(1)	125
C(6)-H(7)	1.32	C(8)-C(6)-C(2)	120
C(6)-C (8)	1.14	H(9)-C(8)-C(6)	121
C(8)-H(9)	1.12	H(10)-C(8)-C(6)	120
C(8)-H(10)	1.12	H(4)-C(1)H(3)	118
		C(6)-C(2)-H(5)	116
		C(8)-C(6)-H(7)	118
		H(10)-C(8)-H(9)	119

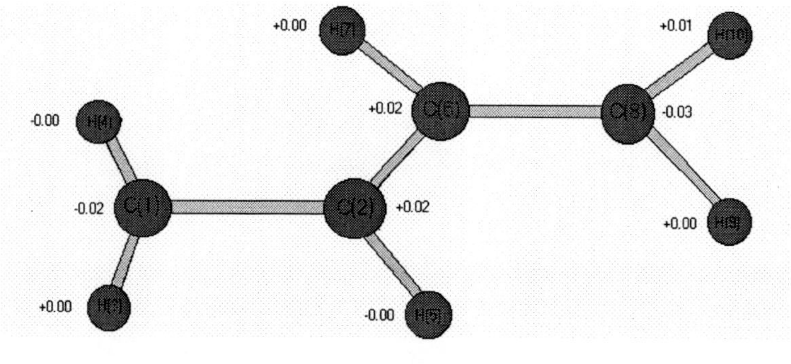

Figure 1. Geometrical and electronic structure butadien-1,3. (E_O = -85.875 kJ/mol, E_b = -11.148 kJ/mol, D = 0.38 dB.).

Table 2. Lengths of bonds and valent corners piperilen

Lengths of bonds	R,A	Valent corners	Degree
C(1)-C(4)	1.34	C(4) -C(1)-H(3)	122
C(1)-H(3)	1.12	C(1)-C(4)-H(5)	120
C(1)-H(2)	1.10	C(1)-C(4)-C(6)	127
C(4)-H(5)	1.12	C(4)-C(6)-H(7)	119
C(4)-C(6)	1.46	C(4)-C(6)-C(8)	123
C(6)-H(7)	1.10	C(6)-C(8)-H(9)	120
C(6)-C(8)	1.34	C(6)-C(8)-C(10)	124
C(8)-H(9)	1.12	C(8)-C(10)-H(11)	112
C(8)-C(10)	1.46	C(8)-C(10)-H(12)	112
C(10)-H(11)	1.12	C(8)-C(10)-H(13)	112
C(10)-H(12)	1.12	H(5)-C(4)-C(6)	113
C(10)-H(13)	1.12	C(8)-C(6)-H(7)	118
		H(9)-C(8)-C(10)	116
		H(2)-C(1)-C(4)	91
		H(2)-C(1)-H(3)	147

Table 3. Lengths of bonds and valent corners dien isoprene

Lengths of bonds	R,A	Valent corners	Degree
C(1)-C(2)	1.34	C(3) -C(1)-C(2)	116
C(1)-H(4)	1.12	H(4)-C(1)-C(2)	124
C(1)-H(5)	1.12	H(5)-C(1)-C(2)	125
C(2)-C(3)	1.46	H(6)-C(3)-C(2)	119
C(3)-H(6)	1.12	H(7)-C(3)-C(2)	110
C(3)-H(7)	1.12	H(8)-C(3)-C(2)	110
C(3)-H(8)	1.12	C(9)-C(2)-C(1)	117
C(2)-C(9)	1.44	H(10)-C(9)-C(2)	122
C(9)-H(10)	1.12	C(11)-C(9)-C(2)	126
C(9)-C(11)	1.33	H(12)-C(11)-C(9)	125
C(11)-H(12)	1.12	H(13)-C(11)C(9)	122
C(11)-H(13)	1.12	H(4)-C(1)-H(5)	111
		C(3)-C(2)-C(9)	127

From tables it is seen, that the maximal charge on atom of hydrogen is q^{H+}_{max} at all investigated diens changes in a range 0.00-+ 0.04. Under the formula (1) we can easily determine values pKa which are in a range +43.7- + 36.1. That is, all researched diens have rather low acid force. Calculations do not cause doubts about what testifies accordingly the following fact, for example. It is known [1], that due to inductive effect of methyling group any isoprene is more active in relation to cations, than butadiene. The data of quantum chemical calculations testify since $q_{c\alpha}$ (butadien-1,3) =-0.02, and $q_{c\alpha}$ (isoprene) =-0.05 also in connection with that a negative charge on a carbon atom $q_{c\alpha}$ criteria of propensity monomers to cationic polymerizations. It is obvious, that if prior to the beginning of polymerization of diens do not change acid force of the monomeasure by introduction of various additives as H-acids the efficiency of process of polymerization will not depend on acid force of olefin.

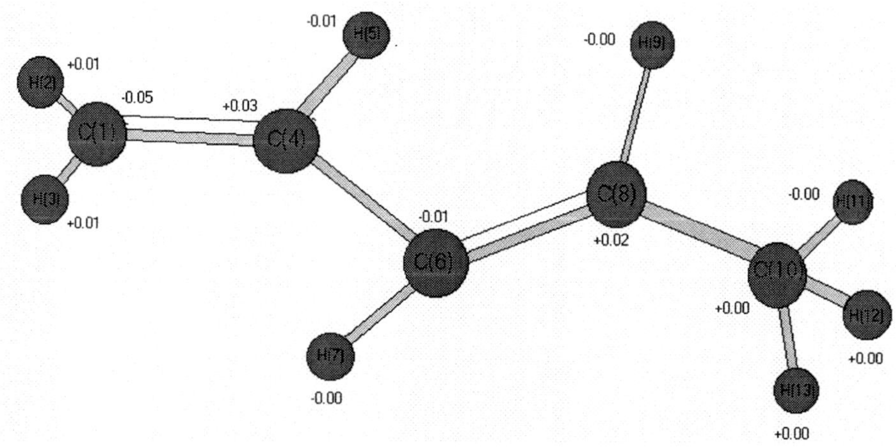

Figure 2. Geometrical and electronic structure piperilen. (E_O = -108.748 kJ/mol, E_b. = -14.505 kJ/mol, D = 0.63 dB.).

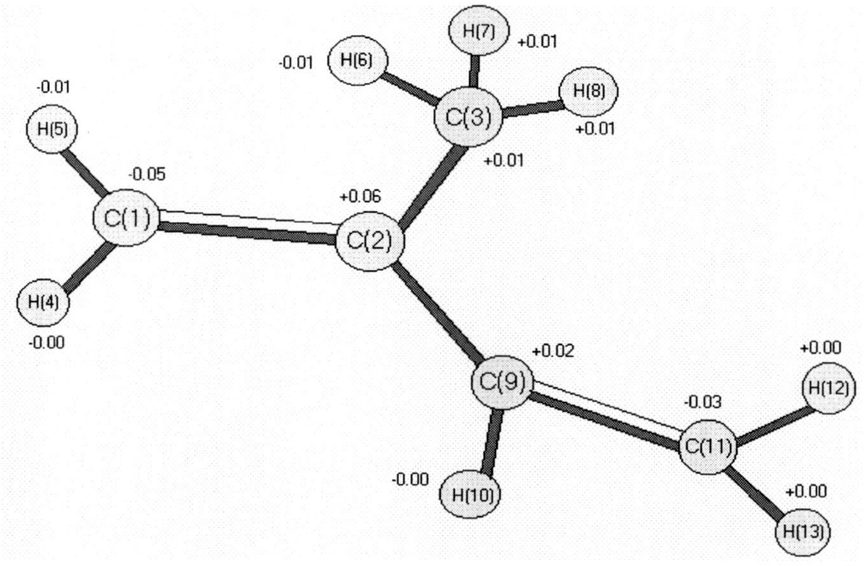

Figure 3. The Geometrical and electronic structure isoprene. (E_O = -108.721 kJ/mol, E_b.. = -14.452 kJ/mol, D = 0.20 dB.).

Table 4. Lengths of bonds and valent corners dien cis, cis - gexadien -2,4

Lengths of bonds	R,A	Valent corners	Degree
C(1)-C(2)	1.34	C(3) -C(1)-C(2)	125
C(1)-C(3)	1.45	C(2)-C(1)-H(4)	122
C(1)-H(4)	1.12	C(1)-C(3)-H(5)	114
C(3)-H(5)	1.12	C(1)-C(3)-H(6)	112
C(3)-H(6)	1.12	C(1)-C(3)-H(7)	112
C(3)-H(7)	1.12	C(1)-C(2)-H(8)	121
C(2)-H(8)	1.12	C(1)-C(2)-C(9)	128
C(2)-C(9)	1.46	H(10)-C(9)-C(2)	118
C(9)-H(10)	1.12	C(2)-C(9)-C(11)	123
C(9)-C(11)	1.35	C(9)-C(11)-H(12)	120
C(11)-H(12)	1.12	C(9)-C(11)-C(13)	123
C(11)-C(13)	1.45	C(11)-C(13)-H(14)	112
C(13)-H(14)	1.12	C(11)-C(13)-H(15)	113
C(13)-H(15)	1.12	C(11)-C(13)-H(16)	113
C(13)-H(16)	1.12	C(3)-C(1)-H(4)	113
		H(8)-C(2)-C(9)	111
		H(10)-C(9)-C(11)	118
		H(12)-C(11)-C(13)	117

Table 5. Lengths of bonds and valent corners dien trans, trans - gexadien-2,4

Lengths of bonds	R,A	Valent corners	Degree
C(1)-C(2)	1.34	H(3) -C(2)-C(1)	120
H(10)-C(9)	1.12	H(4)-C(1)-C(5)	113
C(2)-H(3)	1.12	H(4)-C(1)-C(2)	122
C(1)-H(4)	1.12	C(5)-C(1)-C(2)	125
C(1)-C(5)	1.46	H(6)-C(5)-C(1)	114
C(5)-H(6)	1.12	H(7)-C(5)-C(1)	112
C(5)-H(7)	1.12	H(8)-C(5)-C(1)	112
C(5)-H(8)	1.12	C(9)-C(2)-C(1)	128
C(9)-C(2)	1.43	H(10)-C(9)-C(2)	118
C(11)-C(9)	1.34	H(3)-C(9)-C(2)	114
H(12)-C(11)	1.12	H(12)-C(11)-C(9)	120
C(13)-C(11)	1.46	C(11)-C(9)-C(2)	124
H(14)-C(13)	1.12	H(10)-C(9)-C(11)	112
H(15)-C(13)	1.12	C(13)-C(11)-C(9)	124
H(16)-C(13)	1.12	H(12)-C(11)-C(13)	116
		H(14)-C(13)-C(11)	112
		H(15)-C(13)-C(11)	113
		H(16)-C(13)-C(11)	113

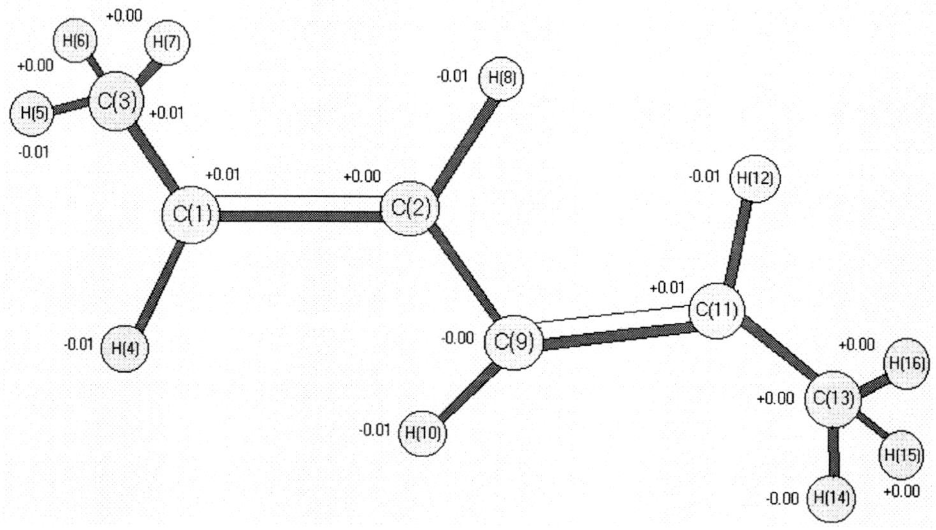

Figure 4. The Geometrical and electronic structure cis, cis - gexadien -2,4. (EO = -131.593 kJ/mol, Eb. = -17.836 kJ/mol, D = 0.17 dB.).

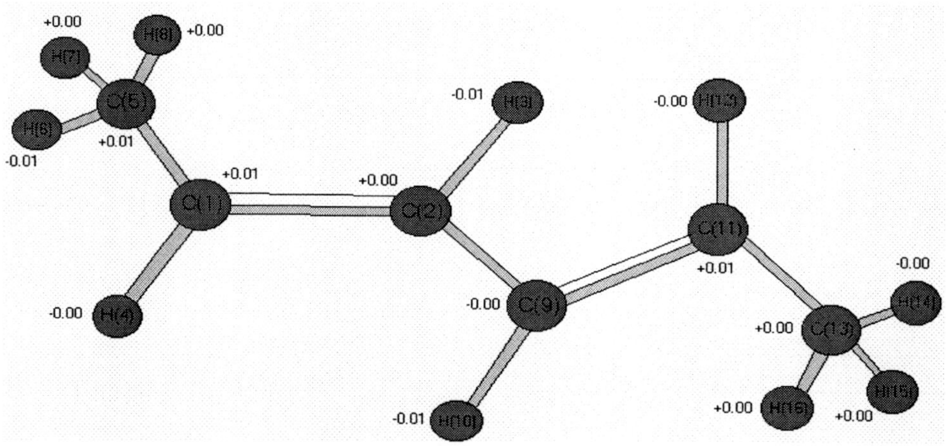

Figure 5. the Geometrical and electronic structure trans, trans - gexadien-2,4. (E_O = -131.593 kJ/mol, $E_{b.}$. = -17.836 kJ/mol, D = 0.17 dB.).

Table 6.. Lengths of bonds and valent corners dien 1,3-dimethylbutadien-1,3

Lengths of bonds	R,A	Valent corners	Degree
C(1)-C(2)	1.46	C(3) -C(1)-C(2)	121
C(1)-C(7)	1.33	H(4)-C(3)-C(1)	113
C(1)-C(3)	1.46	H(5)-C(3)-C(1)	112
C(3)-H(4)	1.12	H(6)-C(3)-C(2)	112
C(3)-H(5)	1.12	C(7)-C(1)-C(3)	117
C(3)-H(6)	1.12	H(8)-C(7)-C(1)	125
C(7)-H(8)	1.12	H(9)-C(7)-C(2)	125
C(7)-H(9)	1.12	H(10)-C(2)-C(1)	112

Table 6. Continued

Lengths of bonds	R,A	Valent corners	Degree
C(2)-H(10)	1.12	C(11)-C(2)-C(1)	128
C(11)-C(2)	1.33	H(12)-C(11)-C(2)	123
H(12)-C(11)	1.12	C(13)-C(11)-C(2)	126
C(13)-C(11)	1.44	H(14)-C(13)-C(11)	114
H(14)-C(13)	1.12	H(15)-C(13)-C(11)	113
H(15)-C(13)	1.12	H(16)-C(13)-C(11)	113
H(16)-C(13)	1.12	C(7)-C(1)-C(2)	126
		H(8)-C(7)-C(9)	110
		H(10)-C(2)-C(11)	120
		H(12)-C(11)-C(13)	111

Table 7. Lengths of bonds and valent corners dien 2,3 dimethylbutadien-1,3

Lengths of bonds	R,A	Valent corners	Degree
C(1)-C(2)	1.46	C(3) -C(1)-C(2)	101
C(1)-C(3)	1.34	C(1)-C(3)-H(4)	121
C(3)-H(4)	1.12	C(1)-C(3)-H(5)	124
C(3)-H(5)	1.14	C(6)-C(1)-C(3)	112
C(1)-C(6)	1.45	C(1)-C(6)-H(7)	117
C(6)-H(7)	1.12	C(1)-C(6)-H(8)	113
C(6)-H(8)	1.12	C(1)-C(6)-H(9)	113
C(6)-H(9)	1.12	C(10)-C(2)-C(1)	121
C(2)-C(10)	1.34	C(2)-C(10)-H(11)	124
C(10)-H(11)	1.12	C(2)-C(10)-H(12)	122
C(10)-H(12)	1.12	C(1)-C(2)-C(13)	113
C(2)-C(13)	1.47	C(2)-C(13)-H(14)	110
C(13)-H(14)	1.12	C(2)-C(13)-H(15)	105
C(13)-H(15)	1.12	C(2)-C(13)-H(16)	105
C(13)-H(16)	1.12	H(4)-C(3)-H(5)	115
		C(6)-C(1)-C(2)	144
		C(10)-C(2)-C(13)	126
		H(11)-C(10)-H(12)	114

Table 8. Lengths of bonds and valent corners dien cis -3-methylpentadien-1,3

Lengths of bonds	R,A	Valent corners	Degree
C(1)-C(2)	1.46	C(3) -C(1)-C(2)	130
C(1)-C(3)	1.33	H(4)-C(3)-C(1)	119
C(1)-H(6)	1.12	H(5)-C(3)-C(1)	124
C(3)-H(4)	1.12	H(6)-C(1)-C(3)	115
C(3)-H(5)	1.12	C(7)-C(2)-C(1)	119

Table 8. (Continued)

Lengths of bonds	R,A	Valent corners	Degree
C(2)-C(7)	1.45	H(8)-C(7)-C(2)	114
C(2)-C(11)	1.34	H(9)-C(7)-C(2)	112
C(7)-H(8)	1.12	H(10)-C(7)-C(2)	112
C(7)-H(9)	1.12	C(11)-C(2)-C(1)	111
C(7)-H(10)	1.12	H(12)-C(11)-C(2)	119
H(12)-C(11)	1.12	C(13)-C(11)-C(2)	127
C(13)-C(11)	1.44	H(14)-C(13)-C(11)	114
H(14)-C(13)	1.12	H(15)-C(13)-C(11)	112
H(15)-C(13)	1.12	H(16)-C(13)-C(11)	113
H(16)-C(13)	1.12	H(6)-C(1)-C(2)	115
		H(4)-C(3)-H(5)	110
		C(7)-C(2)-C(11)	130
		H(12)-C(11)-C(13)	114

Table 9. Lengths of bonds and valent corners dien a trans - 3-methylpentadien-1,3

Lengths of bonds	R,A	Valent corners	Degree
C(1)-C(2)	1.47	C(3) -C(2)-C(1)	127
C(1)-C(3)	1.32	H(4)-C(3)-C(1)	122
C(3)-H(4)	1.12	H(5)-C(3)-C(1)	125
C(3)-H(5)	1.12	H(6)-C(1)-C(3)	117
C(1)-H(6)	1.12	C(7)-C(2)-C(1)	114
C(2)-C(7)	1.46	H(8)-C(7)-C(2)	114
C(7)-H(8)	1.12	H(9)-C(7)-C(2)	113
C(7)-H(9)	1.12	H(10)-C(7)-C(2)	113
C(7)-H(10)	1.12	C(11)-C(2)-C(1)	121
C(11)-C(2)	1.34	H(12)-C(11)-C(2)	118
H(12)-C(11)	1.12	C(13)-C(11)-C(2)	130
C(13)-C(11)	1.44	H(14)-C(13)-C(11)	113
H(14)-C(13)	1.12	H(15)-C(13)-C(11)	111
H(15)-C(13)	1.12	H(16)-C(13)-C(11)	111
H(16)-C(13)	1.12	H(4)-C(3)-H(5)	113
		H(6)-C(1)-C(2)	116
		C(7)-C(2)-C(11)	125
		H(12)-C(11)-C(13)	112

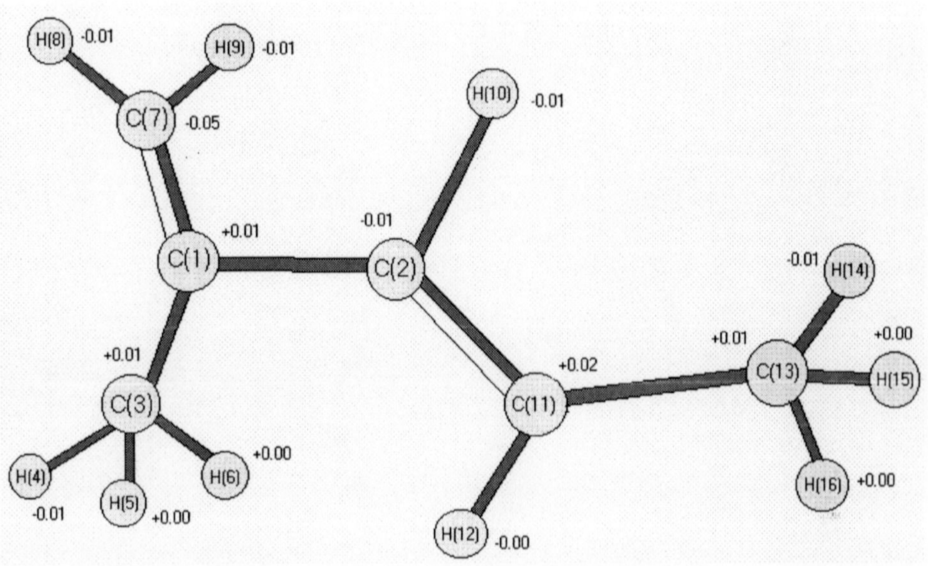

Figure 6. The Geometrical and electronic structure 1,3-dimethylbutadien-1,3. (EO = -131.593 kJ/mol, Eb. = -17.81 kJ/mol, D = 0.86 dB.).

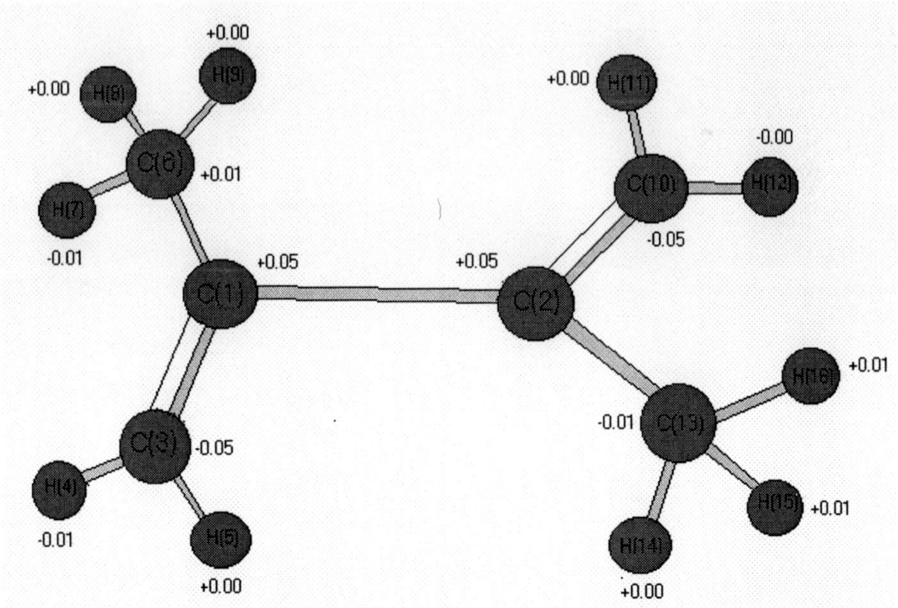

Figure 7. The Geometrical and electronic structure 2,3-dimethylbutadien-1,3. (E_O = -131.515 kJ/mol, $E_{b.}$ = -17.757 kJ/mol, D = 0.13 dB.).

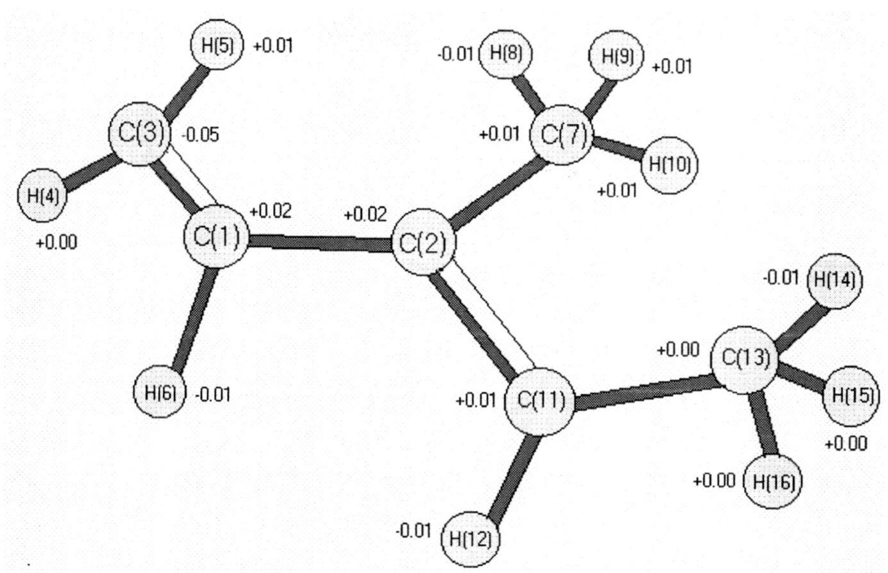

Figure 8. The Geometrical and electronic structure cis -3-methylpentadien-1,3. (E_O = -131.567 kJ/mol, $E_{b.}$ = -17.81 kJ/mol, D = 0.36 dB.).

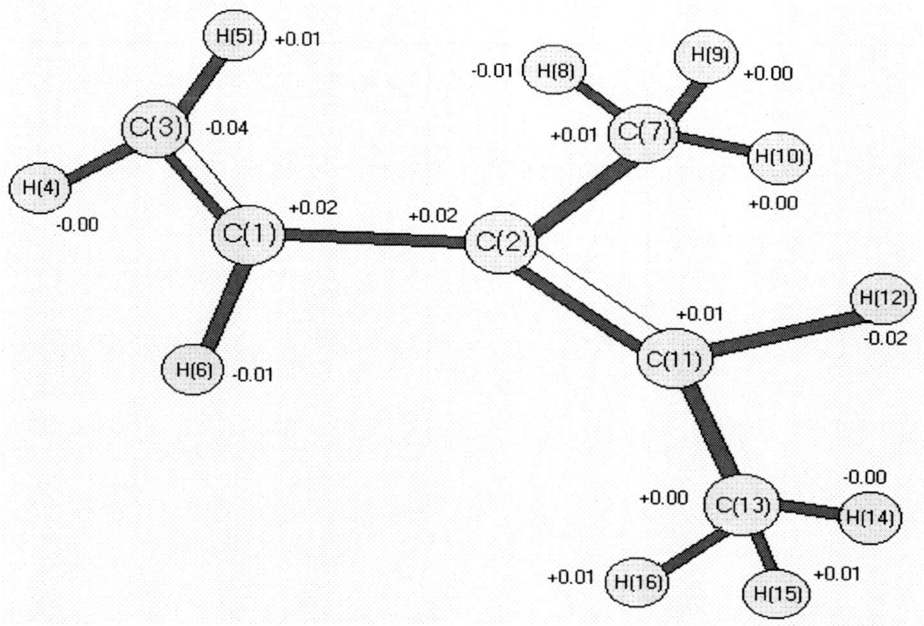

Figure 9. The Geometrical and electronic structure a trance - 3-methylpentadien-1,3. (E_O = -131.593 kJ/mol, $E_{b.}$ = -17.81 kJ/mol, D = 0.3 dB.).

Table 10. Lengths of bonds and valent corners dien 4-methylpentadien

Lengths of bonds	R,A	Valent corners	Degree
C(1)-C(6)	1.44	H(3) -C(2)-C(1)	122
C(1)-C(2)	1.33	H(4)-C(2)-C(1)	123
C(2)-H(3)	1.12	H(5)-C(1)-C(2)	116
C(2)-H(4)	1.12	C(6)-C(1)-C(2)	123
C(1)-H(5)	1.12	C(1)-C(6)-H(7)	118
C(6)-H(7)	1.12	C(1)-C(6)-C(8)	128
C(6)-C(8)	1.34	C(6)-C(8)-C(9)	124
C(8)-C(9)	1.46	H(10)-C(9)-C(8)	113
C(9)-H(10)	1.12	H(11)-C(9)-C(8)	113
C(9)-H(11)	1.12	H(12)-C(9)-C(8)	113
C(9)-H(12)	1.12	C(6)-C(8)-C(13)	121
C(8)-C(13)	1.47	C(8)-C(13)-H(14)	114
C(13)-H(14)	1.12	C(8)-C(13)-H(15)	113
C(13)-H(15)	1.12	C(8)-C(13)-H(16)	113
C(13)-H(16)	1.12	H(3)-C(2)-H(4)	115
		H(5)-C(1)-C(6)	121
		H(7)-C(6)-C(8)	114
		C(9)-C(8)-C(13)	115

Table 11. Lengths of bonds and valent corners dien 1,1,4,4-tetramethylbutadien-1,3

Lengths of bonds	R,A	Valent corners	Degree
C(1)-C(3)	1.46	C(3) -C(1)-C(2)	123
C(1)-H(4)	1.12	H(4)-C(1)-C(2)	122
C(3)-C(5)	1.46	C(3)-C(1)-H(4)	115
C(3)-C(9)	1.46	C(5)-C(3)-C(1)	122
C(5)-H(6)	1.12	H(6)-C(5)-C(3)	101
C(5)-H(7)	1.12	H(7)-C(5)-C(3)	116
C(5)-H(8)	1.12	H(8)-C(5)-C(3)	116
C(9)-H(10)	1.11	C(9)-C(3)-C(1)	126
C(9)-H(11)	1.12	H(10)-C(9)-C(3)	112
C(9)-H(12)	1.12	H(11)-C(9)-C(3)	110
C(1)-C(2)	1.44	H(12)-C(9)-C(3)	110
C(2)-H(13)	1.11	C(5)-C(3)-C(9)	112
C(2)-C(14)	1.34	H(13)-C(2)-C(1)	118
C(14)-C(15)	1.46	C(14)-C(2)-C(1)	131
C(15)-H(16)	1.12	H(13)-C(2)-C(14)	111
C(15)-H(17)	1.12	C(15)-C(14)-C(2)	118
C(15)-H(18)	1.12	H(16)-C(15)-C(14)	113
C(14)-C(19)	1.46	H(17)-C(15)-C(14)	113
C(19)-H(20)	1.12	H(18)-C(15)-C(14)	112
C(19)-H(21)	1.12	C(19)-C(14)-C(2)	122
C(19)-H(22)	1.12	H(20)-C(19)-C(14)	115
		H(21)-C(19)-C(14)	113
		H(22)-C(19)-C(14)	113
		C(15)-C(14)-C(19)	120

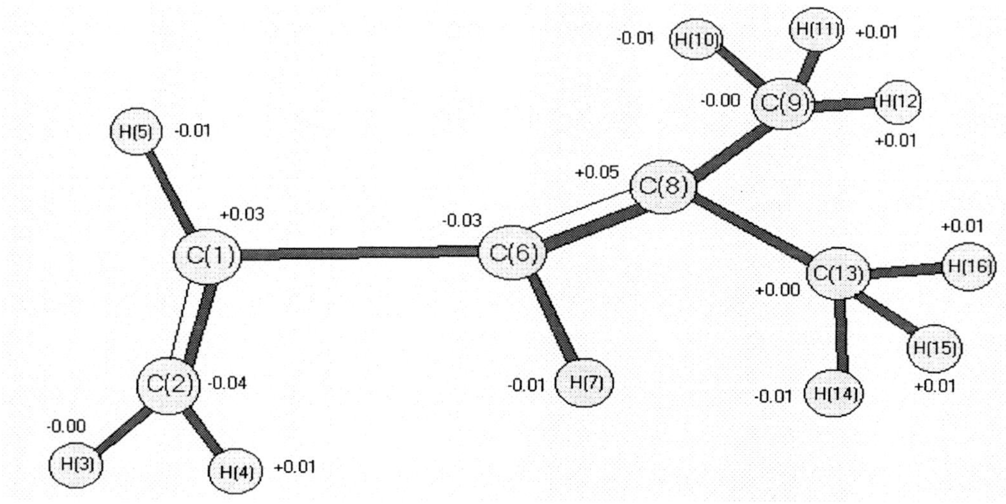

Figure 10. Geometrical and electronic structure 4-methylpentadien. (E_O = -131.567 kJ/mol, $E_{b.}$ = -17.81 kJ/mol, D = 0.73 dB.).

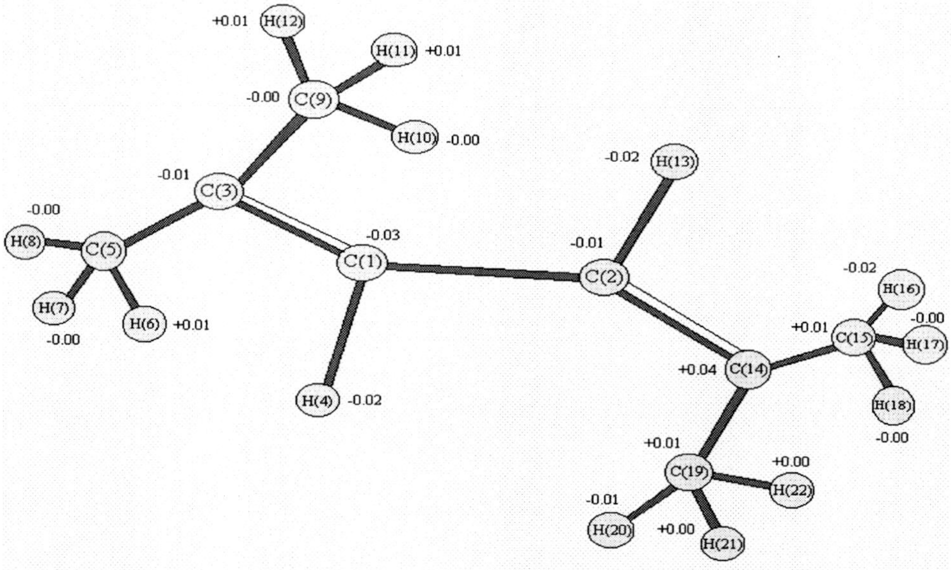

Figure 11. Geometrical and electronic structure 1,1,4,4-tetramethylbutadien-1,3. (EO = -177.233 kJ/mol, Eb. = -24.393 kJ/mol, D = 0.78 dB.).

Table 12. Lengths of bonds and valent corners dien 1,1,3-trimethylbutadien-1,3

Lengths of bonds	R,A	Valent corners	Degree
C(1)-C(6)	1.46	H(3) -C(2)-C(1)	114
C(1)-C(2)	1.46	H(4)-C(2)-C(1)	113
C(2)-H(3)	1.12	H(5)-C(2)-C(1)	113
C(2)-H(4)	1.12	C(6)-C(1)-C(2)	120
C(2)-H(5)	1.12	H(7)-C(6)-C(1)	118
C(6)-H(7)	1.12	H(8)-C(6)-C(1)	114
C(6)-H(8)	1.13	H(9)-C(6)-C(1)	114
C(6)-H(9)	1.13	C(10)-C(1)-C(6)	127
C(1)-C(10)	1.37	H(11)-C(10)-C(1)	122
C(10)-H(11)	1.12	C(12)-C(10)-C(1)	128
C(10)-C(12)	1.49	H(14)-C(13)-C(12)	112
C(12)-C(13)	1.46	H(15)-C(13)-C(12)	113
C(13)-H(14)	1.12	H(16)-C(13)-C(12)	113
C(13)-H(15)	1.12	C(13)-C(12)-(10)	98
C(13)-H(16)	1.12	C(17)-C(12)-C(10)	110
C(12)-C(17)	1.35	H(18)-C(17)-C(12)	119
C(17)-H(18)	1.12	H(19)-C(17)-C(12)	121
C(17)-H(19)	1.10	C(2)-C(1)-C(10)	113
		H(11)-C(10)-C(12)	110
		C(13)-C(12)-C(17)	152
		H(18)-C(17)-H(19)	120

Table 13. Lengths of bonds and valent corners dien cis, trans - gexadien-2,4

Lengths of bonds	R,A	Valent corners	Degree
C(1)-C(2)	1.33	C(3) -C(2)-C(1)	130
H(10)-C(9)	1.12	H(4)-C(3)-C(1)	113
C(2)-C(3)	1.44	H(6)-C(3)-C(1)	112
C(3)-H(4)	1.12	H(7)-C(1)-C(2)	118
C(3)-C(5)	1.12	H(8)-C(2)-C(1)	116
C(3)-H(6)	1.12	C(9)-C(2)-C(1)	129
C(1)-H(7)	1.12	H(10)-C(9)-C(2)	113
C(2)-H(8)	1.12	C(11)-C(9)-C(2)	127
C(9)-C(2)	1.46	H(12)-C(11)-C(9)	117
C(11)-C(9)	1.33	C(13)-C(11)-C(9)	131
H(12)-C(11)	1.12	H(14)-C(13)-C(11)	113
C(13)-C(11)	1.44	H(15)-C(13)-C(11)	111
H(14)-C(13)	1.12	H(16)-C(13)-C(11)	111
H(15)-C(13)	1.12	H(7)-C(1)-C(3)	112
H(16)-C(13)	1.12	H(8)-C(2)-C(9)	116
		H(10)-C(9)-C(11)	120
		H(12)-C(11)-C(13)	112
		H(5)-C(3)-C(1)	112

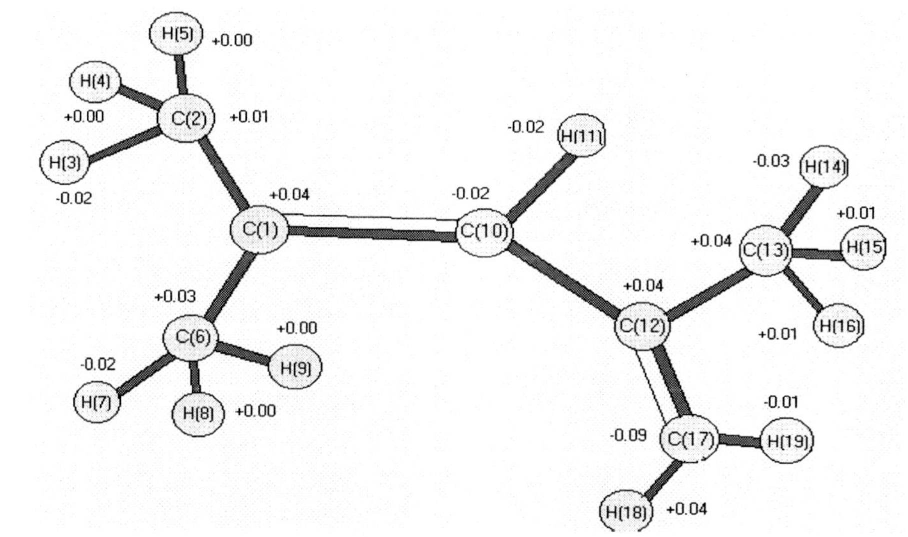

Figure 12. Geometrical and electronic structure 1,1,3-trimethylbutadien-1,3. (EO = -154.282 kJ/mol, Eb.. = -20.984 kJ/mol, D = 1.00 dB.)

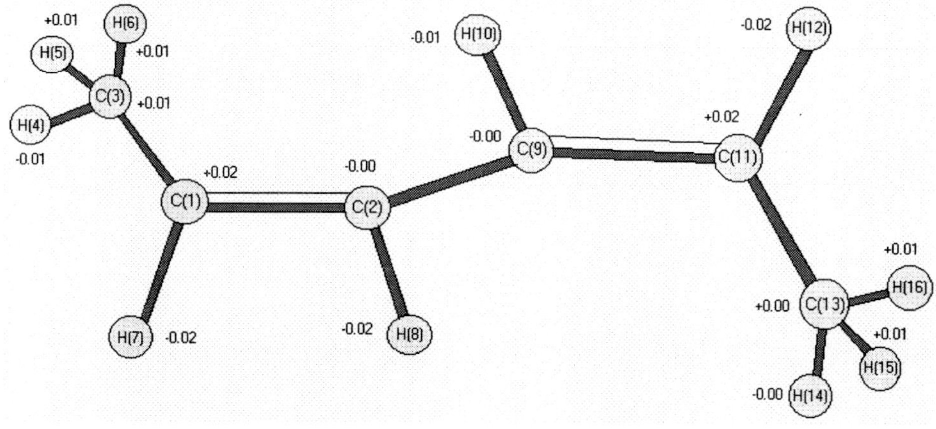

Figure 13. Geometrical and electronic structure cis, trans - gexadien-2,4. (E_0=-131.62 kJ/mol, $E_{b.}$=-17.836A kJ/mol, D=0.19 dB.).

Table 14. Lengths of bonds and valent corners dien 2-ethylbutadien-1,3

Lengths of bonds	R,A	valent corners	Degree
C(1)-C(2)	1.44	C(3) -C(1)-C(2)	129
C(1)-C(3)	1.33	H(4)-C(3)-C(1)	131
C(3)-H(4)	1.12	H(5)-C(3)-C(1)	128
C(3)-H(5)	1.12	H(6)-C(1)-C(3)	126
C(1)-C(6)	1.46	H(7)-C(6)-C(1)	111
C(6)-H(7)	1.12	H(8)-C(6)-C(1)	111
C(6)-H(8)	1.12	C(9)-C(6)-C(1)	111
C(6)-C(9)	1.46	H(10)-C(9)-C(6)	113
C(9)-H(10)	1.12	H(11)-C(9)-C(6)	113
C(9)-H(11)	1.12	H(12)-C(9)-C(6)	113
C(9)-H(12)	1.12	H(13)-C(2)-C(1)	111
C(2)-H(13)	1.15	C(14)-C(2)-C(1)	226
C(14)-C(2)	1.32	H(15)-C(14)-C(2)	123
H(15)-C(14)	1.10	H(16)-C(14)-C(2)	120
H(16)-C(14)	1.10	H(5)-C(3)-H(4)	101
		C(6)-C(1)-C(2)	105
		H(13)-C(2)-C(14)	23
		H(15)-C(14)-H(16)	117

Table 15. Lengths of bonds and valent corners dien 2-tret-buthylbutadien-1,3

Lengths of bonds	R,A	Valent corners	Degree
C(2)-C(1)	1.45	C(3) -C(1)-C(2)	113
C(1)-C(3)	1.34	H(4)-C(3)-C(1)	126
C(1)-C(6)	1.47	H(5)-C(3)-C(1)	127
C(3)-H(4)	1.12	C(6)-C(1)-C(3)	120
C(3)-H(5)	1.12	C(7)-C(6)-C(1)	119
C(6)-C(7)	1.49	H(8)-C(7)-C(6)	112
C(7)-H(8)	1.12	H(9)-C(7)-C(6)	113
C(7)-H(9)	1.12	C(11)-C(6)-C(1)	109
C(7)-H(10)	1.12	H(12)-C(11)-C(6)	113
C(6)-C(11)	1.49	H(13)-C(11)-C(6)	113
C(11)-H(12)	1.12	H(14)-C(11)-C(6)	114
C(11)-H(13)	1.12	C(15)-C(6)-C(1)	109
C(11)-H(14)	1.12	H(16)-C(15)-C(6)	113
C(6)-C(15)	1.49	H(17)-C(15)-C(6)	114
C(15)-H(16)	1.12	H(18)-C(15)-C(6)	112
C(15)-H(17)	1.12	H(19)-C(2)-C(1)	117
C(15)-H(18)	1.12	C(20)-C(2)-C(1)	129
C(2)-H(19)	1.12	H(21)-C(20)-C(2)	123
C(2)-C(20)	1.32	H(22)-C(20)-C(2)	123
C(20)-H(21)	1.12	H(5)-C(3)-H(4)	107
C(20)-H(22)	1.12	C(6)-C(1)-C(2)	127
		H(19)-C(2)-C(20)	114
		H(21)-C(20)-H(22)	114

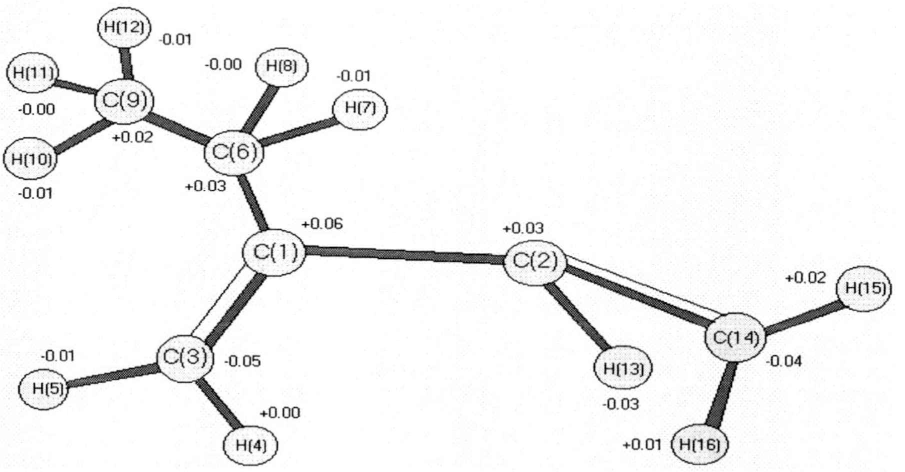

Figure 14. Geometrical and electronic structure dien 2-ethylbutadien-1,3. (E_O = -131.541 kJ/mol, E_b = -17.757 kJ/mol, D = 0.33 dB.).

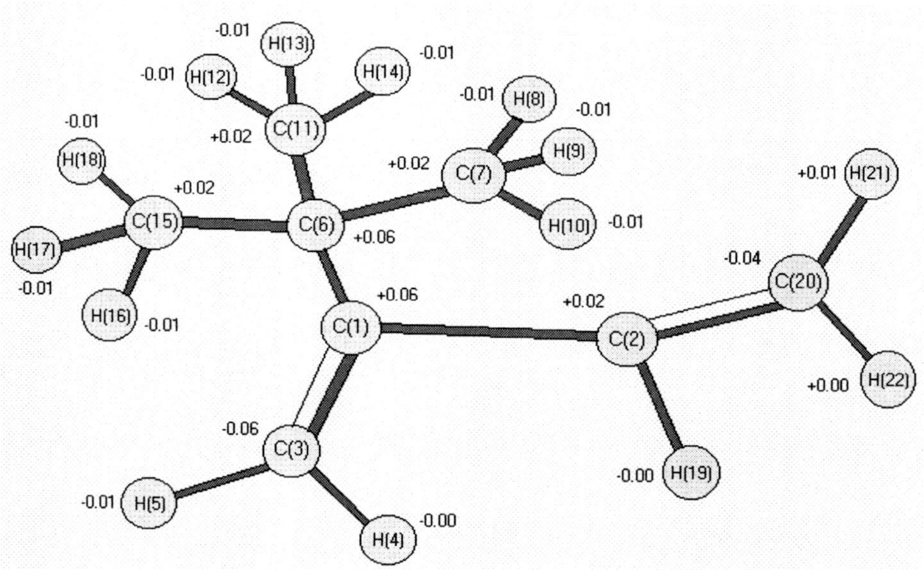

Figure 15. Geometrical and electronic structure 2-tret-buthylbutadien-1,3. (E_O = -177.128 kJ/mol, E_b. = -24.315 kJ/mol, D = 0.31 dB.).

Thus, for the first time us quantum chemical calculation by half empiric method CNDO/2 in parameterization Santry-Popla-Seagala of some typical linear and ramified connected diens was executed: butadien-1,3; piperilen; isoprene; cis, trans - gexadien-2,4; cis, cis - gexadien - 2,4; trans, trance - gexadien-2,4; 1,3- dimethylbutadien-1,3; 2,3-dimethylbutadien-1,3; cis -3-methylpentadien-1,3; trans - 3-methylpentadien-1,3; 4-methylpentadien-1,3; 1,1,3-trimethylbutadien-1,3; 1,1,4,4-tetramethylbutadien-1,3; 2-ethylbutadien-1,3; 2-tret-buthylbutadien-1,3.

Table 16. General energy - E_0, general energy bonds - E_b, dipole moment - D, charges on α and β carbon atoms - $q_{c\alpha}$, $q_{c\beta}$, the maximal charge on atom of hydrogen - q^{H+}, universal parameter of acidity diens - pKa

Diens	E_0, kJ/mol	E_b, kJ/mol	$D_{,dB\cdot}$	$q_{c\alpha}$	$q_{c\alpha 1}$	$q_{c\beta}$	$q_{c\beta 1}$	q_H^+	pKa
1. Butadien -1,3 (C_4H_6)	-86	-11	0.38	-0.02	+0.02	+0.02	+0.01	+0.01	+41.79
2. Piperilen (C_5H_8)	-109	-15	0.63	-0.05	-0.01	+0.03	+0.02	+0.01	+41.79
3. Isoprene (C_5H_8)	-109	-15	0.20	+0.01	+0.12	-0.2	+0.02	+0.01	+41.79
4. cis, trans - gexadien -2,4 (C_6H_{10})	-132	-18	0.19	+0.02	+0.00	+0.00	+0.02	+0.01	+41.79
5. cis, cis - gexadien -2,4 (C_6H_{10})	-132	-18	0.17	+0.01	+0.00	+0.00	+0.01	+0.00	+43.7
6. trans, trans - gexadien -2,4 (C_6H_{10})	-132	-18	0.17	+0.01	+0.00	+0.00	+0.01	+0.00	+43.7
7. 1,3-dimethylbutadien-1,3 (C_6H_{10})	-132	-18	0.86	+0.01	+0.02	-0.01	+0.01	+0.00	+43.7
8. 2,3-dimethylbutadien-1,3 (C_6H_{10})	-132	-18	0.13	-0.05	+0.05	+0.05	-0.05	+0.01	+41.79
9. cis -3-methylpentadien-1,3 (C_6H_{10})	-132	-18	0.36	-0.05	+0.01	+0.02	+0.00	+0.01	+41.79
10. trans - 3-methylpentadien-1,3 (C_6H_{10})	-132	-18	0.3	-0.04	+0.02	+0.02	+0.01	+0.01	+41.79
11. 4-methylpentadien–1,3 (C_6H_{10})	-132	-18	0.73	-0.04	-0.03	+0.03	+0.05	+0.01	+41.79
12. 1,1,3-trimethylbutadien-1.3 (C_7H_{12})	-154	-21	1.00	+0.04	+0.04	-0.02	-0.04	+0.04	+36.1
13. 1,1,4,4-tetramethylbutadien-1,3 (C_8H_{14})	-177	-24	0.78	+0.06	-0.01	-0.03	+0.04	+0.01	+41.79
14. 2-ethylbutadien-1,3 (C_6H_{10})	-132	-18	0.33	+0.06	+0.03	-0.05	-0.04	+0.02	+39.89
15. 2-tret-buthylbutadien-1,3 C_8H_{14})	-177	-24	0..31	+0.06	+0.02	-0.06	-0.04	+0.01	+41.79

It is received optimized by a standard method on all parameters: a geometrical and electronic structure of these connections. Their acid force is theoretically appreciated. It is established, that investigated diens, irrespectively of a degree of branching and a nature dien have acid force and being in a range pKa = + 36.1-43.7 and in this connection they are refered of a class of very weak H-acids (pKa > 14).

REFERENCES

[1] D. J. Kennedy. *Cationic polymerization of olefins*. 1978, Mir Publishers, Moscow, 430 pp.

[2] People I. A., Beveridze D. *Eproximate Molecular Orbit Theory*. 1970, Mc. Grew-Hill, New York, 214 pp.

[3] Babkin V.A. *Structure of the active centers. The mechanism of initiation and growth of a circuit at cation polymerizations olefins at the presence of complex catalysts*. 1999, Moscow, 214 pp. The thesis for degree of Doctor science. Institute of Chemical Physics Russian Academy of Sciences.

In: Preparation and Properties of Monomers, Polymers... ISBN 978-1-60021-557-5
Eds: A. Ballada and G. E. Zaikov, pp. 57-68

Chapter 6

MAGNETIC RECTAL SUPPOSITORIES FOR MEDICAL APPLICATION: INVESTIGATION OF THEIR PHYSICAL AND CHEMICAL PROPERTIES

***O. G. Cherkasova**[*,1]**, S. P. Zavadskiy**[1]**, Yu. Ya. Kharitonov**[1]**, V. A. Frank**[1]**, V. I. Petrov**[1]**, D. G. Krutogin**[2]**, N. A. Brusentsov**[2]* ***and F. S. Bayburtskiy**[3]*

[1]I. M. Sechenov Moscow Medical Academy,
Trubetskaya street, 8 / 2, Moscow 119992, Russian Federation
[2]N. N. Blokhin Russian Cancer Research Center RAMS
Kashyrskoye shosse, 24, Moscow 15478, Russian Federation
[3]N. M. Emanuel Institute of Biochemical Physics, Russian Academy of Sciences,
Kosygina street, 4, Moscow, 119991, Russian Federation

ABSTRACT

The present investigation is devoted to the use of magnetic suppositories for treatment of rectum wounds and after operational fistulas. Physical and chemical properties of used, dispersed magnetic materials in combination with classical medical products are investigated in detail. Opportunities for their safe use in surgical practice are shown.

Keywords: ferrites, magnetic rectal suppositories, treatment of rectum, medical preparations.

INTRODUCTION

At present, the high efficacy and prospects of using the energetic and biotropic properties of magnetic fields and the synergetic action of magnetic fields and medicinal preparations are

* E-mail: Chembio@sky.chph.ras.ru, Bayburt@mail.ru

recognized [1 – 3]. The biotropic properties of magnetic fields are manifested by their antiiflammatory, analgesic, and hypotensive effects and by the positive influence upon hemostasis, metabolism, and regeneration of damaged tissues. These properties are now widely employed in cardiology, surgery, traumatology, dentistry, and other fields of medicine.

Recent developments, already tested *in vivo*, include some new applications such as magnetic fluids, emulsions, and rheological suspensions, magnetic microcapsules and plasters, as well as magnetic rectal capsules and suppositories. In these systems, the magnetic components (fillers) – carriers and sources of the curative constant magnetic field – have the form of finely dispersed powders of ferromagnetic compounds, usually barium ferrite and magnetite. The safety of finely dispersed barium ferrite and magnetite powders was demonstrated in [1, 2]. The magnetic remedies, for example – magnetic rectal suppositories (MRS), may either contain a drug component or not. The curative effect of a drug entering into MRS is usually enhanced under the action of its constant magnetic field.

Despite the obvious success in the study and practical use of magnetic remedies, these products are as yet not provided with exhaustively developed and strictly formulated quality criteria and, hence, not included into pharmacopoeias. Complete normative documentation for these products is not available either. This circumstance hinders the wide use of magnetic remedies in practice. In this paper, we will make an attempt at analyzing, systematizing, and formulating criteria for the quality control of magnetic rectal suppositoria – a highly promising and rather thoroughly studied type of magnetic medicinal remedies – based on the results of our investigations [1 – 14].

According to the classification of magnetic substances [15], barium ferrite belongs to the group of ferromagnetic substances possessing high magnetism, coercive force, and remanent (residual) magnetization, which is a measure of the biotropic properties of a magnetic field. In developing MRS, it is necessary to solve the following tasks: select the magnetic filler, base suppository substance, and drug component; study the compatibility of these components and the stability of their composition; and determine the value of magnetic induction at the MRS surface. In addition, it is necessary to develop methods and techniques for assessing the properties of MRS from the standpoint of both the traditional suppository characteristics and their special magnetic properties. Finally, it is necessary to develop methods for the qualitative and quantitative analysis of MRS components and to characterize MRS with respect to toxicity, microbiological cleanness, and curing effect.

Previously [1 – 3], we have studied the properties of finely dispersed magnetic materials used for the fabrication of MRS and developed physicochemical principles for using these materials in pharmacy. MRS have been successfully tested on a limited number of volunteers in several clinics [1 – 3, 12 – 14].

MATERIALS AND METHODS

Barium Ferrite as a MRS Component

The quality criteria for finely dispersed powder of barium ferrite $BaO \cdot Fe_2O_3$ were formulated in [1 – 3]. MRS can employ domestic commercial products conforming to State

Standards (TU 6-09-1457-67 and TU 6-09-27-254-89). The optimum magnetic properties of barium ferrite powder correspond to the composition BaO · $n$$Fe_2O_3$ with $n = 5.50 - 5.75$. The n value can be determined either from the results of chemical analyses or from the Mössbauer data on the effective magnetic field value H for the most intense partial sextet (conventionally denoted as 12k). The H value determined rom the Mössbauer spectra must fall within 414 ± 4 kOe. According to [3], the powdered barium ferrite for MRS must satisfy the following requirements.

Description, Appearance

Barium ferrite appears as a homogeneous odorless and tasteless powder of yellow- brown color. The powder is attracted by a permanent magnet and is stable on storage in air under normal conditions.

Solubility and IR Spectrum

(A) Barium ferrite is virtually insoluble in water and organic solvents and soluble on heating in aqueous nitric acid (2 : 1); the substance exhibits characteristic reactions for barium and iron (III) cations. (B) The IR absorption spectrum of barium ferrite (pelletized with KBr or suspended in Vaseline oil) in the 400 – 4000 cm^{-1} range displays characteristic broad bands in the region of ≈ 450 and 600 cm^{-1}.

X-Ray Density

The density of barium ferrite determined from the results of X-ray diffraction measurements is $5.27 \cdot 10^3$ kg / m^3.

Magnetic Characteristics

Remanent magnetization, not less than 0.33 T; coercive force, not less than 187 kA / m; magnetic energy density, not less than 17.5 kJ / m^3.

Qualitative Determination

A weighted amount of barium ferrite powder is dissolved in nitric acid (2 : 1) and barium is determined in the form of barium sulfate or by sulfatometric titration with metallochromic indicators (Nitrochromeazo and Orthanilic A). Iron is determined either gravimetrically (in the form of Fe_2O_3), titrimetrically (permanganatometrically, dichromatometrically, iodometrically), or sectrophotometrically. The value of n in the formula BaO · $n$$Fe_2O_3$ must fall within $n = 5.50 - 5.75$. The base substance content in the analyzed powder (with an allowance for the particular n) must be not less than 90 %.

Microbiological Cleanness

The test for microbiological cleanness is conducted in accordance with the State Pharmacopoeia (RSP-XI).

A disadvantage of the commercial samples of barium ferrite, as a component for MRS, is contamination with microorganisms. Experiments showed [13] that this drawback can be eliminated by firing the barium ferrite powder prior to introducing it into the MRS composition. Once this is done, the MRS composition satisfies the RSP-XI requirements with respect to the microbiological cleanness. In addition to ensuring a significant decrease in microbial contamination, the preliminary firing treatment leads to an increase in the remanent magnetic induction due to the improved crystal structure of the barium ferrite powder (heat treatment decreases the weight fraction of a nonmagnetic phase and increases that of the magnetic phase).

The results of our experimental evaluation of the remanent magnetic induction of various magnetic compositions showed that barium ferrite is preferred as a magnetic component for MRS. This compound is chemically stable and nontoxic to the organism when used in the MRS composition [16, 17], while ensuring a relatively high value of the remanent magnetic induction at the MRS surface.

The above criteria for the quality of barium ferrite powder as a special magnetic substance must be met in the stage of preparing this component for the MRS fabrication under commercial or laboratory conditions.

Results and Discussion

Special Features of the Quality Criteria for MRS

We have developed the technology (laboratory scale) and principal quality criteria for the MRS of parmidin [1 – 3, 12], novocain [13], and ortophen [14] with barium ferrite as the magnetic filler component. The special features of evaluating the quality of such MRS consist in the necessity of checking their properties from the standpoint of both the traditional suppositoria characteristics and special magnetic parameters.

According to RSP-XI, traditional rectal suppositoria must obey certain requirements with respect to the following properties: homogeneity, weight, dimensions, melting temperature, solidification temperature, and total deformation time (for suppositoria on a hydrophobic base) or dissolution time (for suppositoria on a hydrophilic base). In addition, the traditional suppositoria are characterized by several parameters of the base, including acid number (AN), iodine number (IN), peroxide number (PN), microbiological cleanness, pH of the aqueous extract, bioaccessibility of a drug in the composition, and toxicity. All these requirements apply to MRS as well.

As indicated above, the special parameters characterizing the magnetic properties of MRS include the remanent magnetic induction (B) at the suppository surface, the spatial distribution of the magnetic field strength near this surface (magnetic field topography), the identity of a magnetic filler substance, and the quantitative content of this magnetic filler in the MRS composition.

The remanent magnetic induction at the MRS surface is determined by a conventional ballistic method (e.g., with a milliteslaammeter of the F – 4354 / 1 type). Homogeneity of the spatial distribution of the magnetic field is assessed and the field strength near the MRS poles is determined using a magnetooptical compensation method with photoregistration.

The principle of this method consists in measuring the size of a single-domain region formed in the labyrinth structure of a $(BiY)_3(FeJa)_5O_{12}$ magnetooptical film under the action of the MRS dissipated magnetic field component perpendicular to the film surface. Using a setup comprising the garnet film, a Helmholtz coil with a field direction opposite to that of the MRS field, controlled dc power supply, and an ammeter, it is possible to determine the magnetic field strength (by measuring the coil current) at various distances from the MRS surface. The magnetooptical-film data obtained for MRS samples with different values of the surface induction showed that the magnetic field strength drops sharply with increasing distance from the MRS poles in a manner that is virtually independent of the character of magnetization.

There is a linear relationship between the magnetic induction at the MRS surface and the depth of the magnetic field penetration into the biological object.

Special Features of MRS Technology

It was established [13, 14] that both the remanent magnetic induction at the MRS surface and the spatial distribution of the magnetic field strength depend on the weight fraction of the magnetic filler in the composition, on the texturing – a special feature of the MRS technology related to the oriented distribution of the magnetic filler in the suppository, and on the character of MRS magnetization (longitudinal, transverse, single- or multipole). The texturing was achieved by magnetizing MRS between the poles of an electric magnet producing a sufficiently homogeneous field with a strength of 1200 kA / m, where the MRS were exposed with their long axes oriented along the field direction.

In order to control the biotropic properties of MRS, it is necessary to establish a relationship between the magnetic filler characteristics, its weight fraction, and the presence of a magnetic and crystal texture in the suppository body, the character of magnetization, and the therapeutic properties of barium ferrite MRS.

Previously [13], we determined the optimum conditions for the fabrication of MRS with transverse and longitudinal magnetization, measured the patterns of magnetic field distribution, determined the optimum texturing regime, and studied the influence of magnetic filler on the drug component activity, microbiological cleanness, and stability on storage. We have also studied the possibility of increasing the properties of the magnetic filler. Optimum characteristics are only provided by strictly following the corresponding conditions in all technological stages.

The stages of MRS fabrication technology are generally the same as those for the traditional products. Distinctive features of the MRS fabrication include the stage of suppository mass dosing and the stage of suppository shaping until complete solidification in a homogeneous magnetic field (anisotropic suppositoria). The products not processed in the last stage are called isotropic (these MRS are characterized by unoriented magnetic filler distribution). Based on the experimental data, we did not establish significant differences in the vales of magnetic induction at the surface of MRS magnetized under various conditions (transverse versus longitudinal).

An optimum texturing regime corresponds to a process time of 5 min and an orienting field strength of 300 kA / m. An increase in the exposure duration or the applied field strength does not lead to a significant increase in the remanent magnetic induction.

Magnetooptical Method for Determining the Spatial Distribution and Strength of a Constant Magnetic Field Near MRS (Magnetic Field Topography)

Using the magnetooptical method, it is possible to assess the homogeneity of the magnetic field around MRS and measure themagnetic field strength at various points in the space. Proceeding from the fact that the magnetic field penetration increases with the magnetic filler weight fraction in MRS, we studied products with a high barium ferrite fraction, varying from 50 to 70 wt.%. The maximum range of the magnetic field was observed for MRS with a barium ferrite weight fraction of 70 % and reached 5.83 mm.

Prospects for Increasing the Magnetic and Microbiological Properties of the Magnetic Filler

The magnetic properties of barium ferrite MRS depend primarily on the weight fraction of barium ferrite in the composition and on the presence of texture in the suppository body. The optimum powder preparation technology for hydrophobic MRS consists in dry milling of barium ferrite followed by firing for 2 h at 950°C. The additional firing treatment of barium ferrite increases the degree of ferritization (the weight fraction of magnetic phase) and the magnetic induction and provides for the required microbiological cleanness of the magnetic component in MRS. Following all these stages, the magnetic filler powder conforms to the RSP-XI requirements with respect to microbiological cleanness and can be used in the MRS composition without additional conservants.

Double-Layer Magnetic Rectal Suppositories

Besides the aforementioned advantages of MRS as convenient and comfortable sources of a constant magnetic field, the final products have certain drawbacks, including their black color determined by the presence of the magnetic filler (barium ferrite) and rapid variation in the shape of the magnetic field source (and, hence, of the magnetic properties of MRS) upon introduction into rectum. In order to remove these drawbacks, it was suggested to use double-layer MRS (DMRS). DMRS are fabricated with ortophen as the drug component.

DMRS are composed of two layers. The inner layer is a relatively high-melting composition of black color, which contains the magnetic filler without drug component. The outer layer is a low-melting composition of white or light-yellow color containing only the drug and no magnetic component. These DMRS were manufactured and quality assessed [14] in accordance with all the general and special criteria formulated above. The IR spectroscopic investigation showed that the DMRS components are fully compatible.

Selecting Suppository Base for MRS

As mentioned above, the primary task in developing an MRS technology is to select a suppository base component. The requirements to this component are as follows: rated room-temperature hardness, ability to melt or dissolve at human body temperature, chemical and pharmacological inertness, ability to easily release the drug component, the absence of irritant properties, and stability with respect to light, moisture, oxygen from air, and microorganisms.

Table 1. Melting points of bases for external DMRS layers*

Base composition	$Tm \pm \Delta T$, °C
Witepsol W_{35} – witepsol H_{15} (50 : 50)	35.0 ± 1.0
Witepsol W_{35} – witepsol H_{15} (33.5 : 66.5)	35.5 ± 1.0
Witepsol W_{35} – witepsol H_{15} (25 : 75)	36.0 ± 1.0
Type A solid fat – T1 emulsifier (95 : 5)	33.5 ± 1.0
Type A solid fat – T1 emilsifier – gelatin – water distilled (84.8 : 5 : 1.3 :8.9)	33.5 ± 1.0

Table 2. Melting points of bases for internal DMRS layers

Base composition	$Tm \pm \Delta T$, °C
Paraffin – witepsol H_{15} (50 : 50)	38.4 ± 1.0
Paraffin – witepsol H_{15} (30 : 70)	37.0 ± 1.0
Paraffin – witepsol H_{15} (40 : 60)	39.01 ± 1.0
Paraffin – witepsol H_{15} (30 : 70)	37.7 ± 1.0
Paraffin – witepsol H_{15} (20 : 80)	36.9 ± 1.0

*Here and below data represent either confidence intervals with a confidence probability of $\alpha = 0.95$ or average values over the results of n measurements.

Studied as possible suppository bases were witepsol, salomas (hydrogenated fat), and cacao oil, which obey all the above requirements. Therefore, these substances can be recommended as suppository base materials for MRS.

Creating DMRS posed the task of selecting additional bases. As noted above, the DMRS comprise a low-melting outer layer with the drug and a high-melting inner layer with the magnetic filler. Accordingly, the outer and inner layers requires using base materials with lower (Table 1) and higher (Table 2) melting points. Data presented in Tables 1 and 2 show that the proposed substances are compatible with the other DMRS components, do not affect the remanent magnetic induction level, and hence, can be used as base components for DMRS.

Evaluation of the Physicochemical Properties of MRS with Drugs

Quality Evaluation of MRS with Parmidin

The MRS with parmidin [1, 12] were fabricated using finely dispersed barium ferrite as the magnetic filler, cacao oil, witepsol, salomas (an alloy containing 5% of T1 emulsifier and 95% of Type A solid fat), and a poly(ethylene oxide) mixture (95% PEO-1500 and 5% PEO-400) as the suppository bases. An analysis of the IR absorption spectra showed that all these components in MRS compositions are compatible. The equilibrium dialysis data showed a decrease in the rate of parmidin release from the compositions with barium ferrite. This was explained by the increase in the system viscosity as a result of introduction of the magnetic component.

A special technology developed for the fabrication of MRS includes the stage of suppository weight dosing and the stage of suppository shaping until complete solidification in a sufficiently homogeneous magnetic field, followed by magnetization of the MRS and measurement of the magnetic induction at the suppository poles. A test for stability of the MRS stored for 18 months at a temperature 5°C in a refrigerator showed that both composition and properties of the MRS remained unchanged.

The quantitative analysis for parmidin in MRS is performed by spectrophotometry at λ = 264 nm (relative to water) using a 0.1 mole / liter solution. This technique is compatible wit the presence of a magnetic filler, which affects neither accuracy not reproducibility of analyses. Data on the main quality parameters and their stability on storage for MRS with parmidin are listed in Table 3.

Table 3. Quantitative data on the composition stability of MRS with parmidin based on salomas (95 % type a solid fat – 5 % T1 emulsifier) stored at 5°C (Average for *n* measurements)*

Characteristic	Storage time, months				
	0	3	6	12	18
TDT, min (n = 3)	40.0	40.5	40.5	41.0	41.2
Hardness P, g (n = 5)	3000	3200	3100	3700	4400
pH (n = 3)	7.0	6.9	7.2	7.4	7.2
Tm,°C (n = 3)	34.2	34.3	–	34.5	34.3
Tsol,°C (n = 3)	30.0	–	–	30.0	30.0
AN (n = 3)	0.42	–	0.45	0.45	0.44
PN (n = 3)	0.03	0.03	0.03	0.03	0.025
m (parmidin) ± Δm, g (α = 0.95, n = 5)	0.245 ± 0.001	0.2447 ± 0.0008	0,2442 ± 0.0006	0.245 ± 0.001	0.240 ± 0.001
m (MRS), g (n = 10)	4.25	4.25	4.30	4.25	4.20
B, T (n = 5)	46	46	45	45	45

* Here and Tables 4 – 6: P = hardness (g); m = mass (g), B = magnetic induction (mT); TDT = total deformation time; Tm = melting point (°C); IN = iodine number; AN = acid number; PN = perooxide number.

Quality Evaluation of MRS with Novocain

The MRS with novocain [13] were fabricated using finely dispersed barium ferrite as the magnetic filler and witepsol H-15 as the suppository base.

MRS with novocain were fabricated in the state of longitudinal magnetization. Data of the IR and UV spectrophotometry, potentiometry, titrimetry, and magnetic measurements confirmed the compatibility of novocain with all other MRS components. The content of novocain in witepsol-based MRS can be determined by any of three methods (titrimetric, potentiometric, and spectrophotometric) [13]. It was established that the composition and properties of MRS with novocain remain unchanged during 24-month storage at a temperature 5°C (Table 4).

Table 4. Quantitative data [13] on the composition stability of MRS with Novocain based on Witepsol H – 15 Stored at 5°C (average for *n* measurements)*

Characteristic	Storage time, months		
	0	12	24
TDT, min (n = 3)	10 ± 1	10 ± 2	11 ± 2
pH ± Δ pH	7.00 ± 0.01	7.00 ± 0.06	6.95 ± 0.07
Tm, °C	34 ± 2	34 ± 3	34 ± 2
IN ± Δ IN	5.84 ± 0.03	5.62 ± 0.04	5.76 ± 0.03
AN ± Δ AN	0.13 ± 0.04	0.19 ± 0.06	0.15 ± 0.04
PN ± Δ PN	0.051 ± 0.001	0.067 ± 0.001	0.056 ± 0.002
m (novocain) ± Δm, g	0.096 ± 0.001	–	–
m (MRS), g (n = 10)	4.10 ± 0.03	4.20 ± 0.01	4.15 ± 0.02
B ± Δ B, T (n = 5)	21 ± 2	20 ± 1	19 ± 3

Quality Evaluation of MRS with Ortophen

The MRS with ortophen [14] were fabricated using finely dispersed barium ferrite as the magnetic filler and either a salomas alloy (containing 5% of T1 emulsifier and 95% of type A solid fat) or a witepsol mixture (witepsols H15 and W35 in a 66.5 : 33.5 ratio). The MRS with ortophen were fabricated and assessed with respect to quality similarly to the analogous products with parmidin and novocain. The IR spectra confirmed the compatibility of the base substances with the magnetic filler (barium ferrite) and the drug (ortophen).

In order to determine the optimum weight faction of barium ferrite ensuring a magnetic induction value of 20 – 25 mT near magnetic poles at the suppository surface, MRS samples with both isotropic and anisotropic magnetization were fabricated using the two selected bases and a barium ferrite weight fraction varied from 10 to 70%. Further increase in the barium ferrite content was impossible because MRS became brittle. It was found [14] that MRS with ortophen possessing the required magnetic induction level can be obtained using a 50% weight fraction of barium ferrite in the MRS with anisotropic transverse magnetization and a 60% weight fraction, in isotropic transversely magnetized MRS.

The quantitative content of ortophen in MRS was determined using a specially developed spectrophotometric procedure [14] taking into account the presence of a magnetic filler. It was established that the composition and properties of MRS and DMRS with ortophen remain unchanged during 18-month storage at a temperature 5°C (Tables 5 and 6).

Proposed Quality Criteria for MRS

Based on the results of investigations of the general and special (magnetic) properties of MRS, we suggest using the following special criteria in characterizing the MRS quality: (I) the identity of the magnetic filler; (II) the value of remanent magnetic induction at the suppository surface; (III) the character of spatial distribution and the magnitude of the magnetic field near the suppository; (IV) qualitative analysis of the magnetic filler.

Naturally, all the general quality criteria stipulated for traditional nonmagnetic suppositories apply to the MRS as well.

Table 5. Quantitative data [14] on the composition stability of MRS with Ortopen stored at 5°C (Average for *n* measurements)*

Characteristic	Storage time, months		
	0	12	24
TDT, min (n = 3)	10 ± 3	10 ± 2	9 ± 2
pH ± Δ pH	7.00 ± 0.03	7.05 ± 0.05	7.05 ± 0.05
Tm, °C	33 ± 1	34 ± 3	34 ± 2
AN ± Δ AN	0.65 ± 0.09	0.64 ± 0.07	0.61 ± 0.08
PN ± Δ PN	0.059 ± 0.001	0.053 ± 0.001	0.051 ± 0.002
m (ortophen) ± Δm, g	0.05 ± 0.0003	0.05 ± 0.0004	0.05 ± 0.000
m (MRS), g (n = 10)	4.20 ± 0.05	4.15 ± 0.05	4.15 ± 0.03
B ± Δ B, T (n = 5)	23 ± 2	21 ± 2	20 ± 1

Table 6. Quantitative data [14] on the composition stability of DRMS with Ortophen stored at 5°C (Average for *n* measurements)

Characteristic	Storage time, months		
	0	12	24
TDT, min (n = 3)	12 ± 1	12 ± 1	10 ± 1
pH ± Δ pH	7.00 ± 0.04	7.02 ± 0.08	7.02 ± 0.003
Tm ± ΔT, °C (external layer)	36.0 ± 1.0	37.0 ± 1.0	37.2 ± 1.0
Tm ± ΔT, °C (internal layer)	38.2 ± 1.0	39.0 ± 1.0	38.2 ± 1.0
AN ± Δ AN	0.71 ± 0.08	0.65 ± 0.05	0.61 ± 0.07
PN ± Δ PN	0.060 ± 0.002	0.061 ± 0.002	0.068 ± 0.001
m (ortophen) ± Δm, g	0.05 ± 0.0003	0.05 ± 0.0005	0.05 ± 0.0005
m (MRS), g (n = 10)	4.30 ± 0.05	4.25 ± 0.05	4.25 ± 0.05
B ± Δ B, T (n = 5)	21 ± 2	20 ± 1	20 ± 1

Therapeutic Properties of MRS

The results of a comparative evaluation of the resorptive antiinflammatory effects of traditional (nonmagnetic) and magnetic rectal suppositories with parmidin and the MRS without drugs showed evidence of the more pronounced and prolonged parmidin action in MRS as compared to the drug action in the nonmagnetic suppository. It was established that MRS produces an additional positive local effect: the magnetic field of the suppository increases local blood circulation. Study of the acute and chronic toxicity of MRS with parmidin and those without drugs showed that both types are biologically safe for long-term therapy.

The experiments with MRS containing 0.25 g of parmidin and those without drugs (barium ferrite alone) conducted for 2.5 years showed that MRS produced spasmolytic, analgesic, antiiflammatory, deaggregative, and trophotropic action in patients suffering from inflammatory disorders in the distal part of the rectum, functional disorders of the anus sphincter, and post-operation scarry strictures in the stage of reversible damage. The administration of MRS has proved to be expedient in cases of proctologic disorders such as hemorrhoids, anal fissures, proctitis, proctosigmoiditis, cryptitis, and rectitis. The use of drug-free MRS with a remanent magnetic induction of ≈ 40 mT at the suppository surface in patients suffering from chronic constipation restored the ability for self-defecation after the suppository was used for 3 – 4 days.

Conclusion

The results of experiments *in vitro* and *in vivo* showed evidence that MRS potentiate the local anesthetic effect of novocain as compared to that observed for the drug in traditional (nonmagnetic) suppositories. The most pronounced local anesthetic action was observed for MRS with a magnetic induction of ≈ 20 mT at the magnetic poles. The experiments *in vivo* confirmed the expediency of using the antiinflammatory potential of DMRS and MRS with ortophen with both uniform (MRS) and layered (DMRS) distribution of drug and magnetic filler in the suppository body.

At the same time, the results of tests using MRS with the drugs mentioned above suggested the following counterindications and precautions for the MRS therapy. Absolute counterindications: oncologic processes, pregnancy, presence of magnetic objects in the organism, individual intolerance, acute myocardial infarction, acute insult. Special precautions have to be taken in patients with hypertension in stage III, thyrotoxicosis, acute hemorrhagic bleeding, and systemic blood disorders.

References

[1] O. G. Cherkasova, *Doctoral (Pharm. Sci.) Thesis* [in Russian], Moscow (1993).

[2] Yu. Ya. Kharitonov, O. G. Cherkasova. // *Farmatsiya*, 5 (1996) 14–18.

[3] O. G. Cherkasova, D. G. Krutogin, Yu. Ya. Kharitonov, *Effect of Electromagnetic Fields on the Human Organism* [in Russian], New Millenium Foundation, Moscow (1998), pp. 120–142.

[4] O. G. Cherkasova. // *Khim.-Farm. Zh.*, 26 (7) (1992) 84–88.

[5] O. G. Cherkasova, N. B. Saidov, Yu. Ya. Kharitonov. // *Khim.-Farm. Zh.*, 26 (11) (1992) 93–95.

[6] O. G. Cherkasova, Yu. Ya. Kharitonov, N. B. Saidov. // *Farmatsiya*, 2 (1993)13–16.

[7] D. G. Krutogin, O. G. Cherkasova, M. N. Denisova // *Magnitologiya*, 1 (1994)12–13.

[8] O. G. Cherkasova, G. A. Grigor'eva, Yu. Ya. Kharitonov // *Magnitologiya*, 1 (1994) 39–41.

[9] V. Bykov, V. I. Nikolaev, C. Diaz. // *Hyperfine Interact.*, 67 (1991) 607–609.

[10] O. G. Cherkasova, Yu. Ya. Kharitonov, M. N. Denisova. // *Khim.-Farm. Zh.*, 29 (12) (1995) 49–52.

[11] M. N. Denisova, O. G. Cherkasova, Yu. Ya. Kharitonov, *Farmatsiya*, No. 3, 22–29 (1996).
[12] N. B. Saidov, *Candidate (Pharm. Sci.) Thesis* [in Russian], Moscow (1991).
[13] M. N. Denisova, *Candidate (Pharm. Sci.) Thesis* [in Russian], Moscow (1997).
[14] M. V. Smolyaninova, *Candidate (Pharm. Sci.) Thesis*, Moscow (2000).
[15] N. Remizov, *Medicinal and Biological Physics* [in Russian], Vysshaya Shkola, Moscow (1996), pp. 298–302.
[16] V. G. Karavaev, E. V. Ivashkevich, Zh. V. Martynenko. // *Vestn. Rentgenol.* 1 (1978) 36–40.
[17] M. Granov, V. Yu. Derkach, V. Kh. Saikin. // *Abstracts of Papers. The 2nd Regional Sci.-Pract. Conf.* [in Russian], Rostov-on-Don (1985), p. 137.

In: Preparation and Properties of Monomers, Polymers... ISBN 978-1-60021-557-5
Eds: A. Ballada and G. E. Zaikov, pp. 69-74

Chapter 7

Studying of a Magnetic Resonance in Contrasting Agents on the Basis of Biodecomposed Magnetic Fluids

***F. S. Bayburtskiy**[*1], **G. M. Semenova**[1], **N. A. Brusentsov**[2], **M. I. Papisov**[2], **A. Yu. Bogdanov**[2], **A. N. Bogdanova**[2], **E. Yu. Filinova**[2] and **A. L. Nikolaev**[3]*

[1]N. M. Emanuel Institute of Biochemical Physics, Russian Academy of Sciences
4, Kosygin str., Moscow 119991, Russian Federation
[2]N. N. Blokhin Russian Cancer Research Center RAMS
24, Kashyrskoye shosse, Moscow 15478, Russian Federation
[3]M. V. Lomonosov Moscow State University, Department of Chemistry, Sub-faculty of radiochemistry, Vorobyovy gory, Moscow 119899, Russian Federation

Abstract

In the present work the magnetic resonance in contrast agents has been investigated. Contrast agents represented biodecomposed magnetic fluids. Surface structure of dextran-coated superparamagnetic particles is studied as a biodistribution-defining factor. Colloidal particle and polymer molecule of the same size and similar outer structure are shown to possess similar biokinetics and biodistribution. A dense dextran layer on the particle surface is identified as a structure responsible for particle accumulation in the lymph nodes.

Keywords: magnetic fluids, biokinetics and biodistribution, superparamagnetic particles.

---

[*]E-mail: Chembio@sky.chph.ras.ru, Bayburt@mail.ru

Introduction

Because of the intrinsic differences in relaxation times, T1 and T2, in normal and pathologic tissue, magnetic resonance (MR) imaging has demonstrated excellent diagnostic sensitivity. Various magnetic pharmaceuticals (MR contrast agents), which modify proton relaxation parameters, have been developed to increase both diagnostic sensitivity and specificity. Typically, the distribution of these agents to normal and pathologic tissue is different resulting in increased contrast and thus improved detection of disease. The more specific the distribution of an agent to one tissue without distribution to its surrounding, the more effective the agent.

An ideal magnetic pharmaceutical is one that can be detected by MR at very low concentrations (<< 1 mg / l). Colloidal iron oxide particles, which generate high magnetic field gradients near the surface (microinhomogeneities), meet this detection requirement [1].

Colloidal superparamagnetic preparations are widely studied as prototypes of a targeted contrast agents for MR imaging containing antibodies or receptor ligands attached to particle surface [2]. These high-affinity molecules bind to certain components of the target tissue site, allowing agent accumulation at this site. However, in vivo kinetics and biodistribution of the particles are often altered by blood protein adsorption to particles (opsonization) or by direct cellular recognition. These processes are associated with rapid phagocytosis resulting in distribution of the material mainly to the liver and spleen. In this study, we investigate the surface structure and the effects of surface modification on the biokinetics of superparamagnetic dextran-coated nanoparticles.

Materials and Methods

The dextran-coated monocrystalline iron oxide nanoparticles (NION-37 and MION-46) were prepared by magnetite synthesis in presence of dextran (molecular weight 10 000 Dalton); generally following the synthatic rout described in article [3] with modifications providing different particle properties. Particles of MION-37 were formed using Fe (III) / Fe (II) ratio 2 / 1, with subsequent oxidation by air oxygen. MION-46 was synthesized using high (35 %) concentration of dextran, low (150 mM) concentration of iron and Fe (II) / Fe (III) = 1 / 2 at 5°C, followed by incubation at 70°C. A radioactive label (^{111}In) was entrapped into the core crystal structure by coprecipitation [4]. The iron concentration in the particle preparations were determined by spectrophotometry at 430 nm: the sample contained 2 ml 11 M HCl. 20 μl ferrofluid and 20 μl 3 % H_2O_2. The concentration of dextran was determined by the phenolsulphate method [5].

The particle core diameter was measured by electron transmission microscopy. The total size of particle in water was measured as hydrated diameter by laser light scattering (LLS) and gel chromatography following conventional techniques.

Particle-conatining reverse micelles were prepared in water / octan / Aerosol OT (W / O / AOT = 1 / 50 / 0.5) system by mixing colloidal solution of particles with AOT solution in octane. To control the accuracy of measurements, standard 94 nm latex was solubilized in the samples of particle containing micelles following the same technique. Dehydration of

Sephadex G-200 beads was studied under the same conditions; bead size was controlled by optical microscopy.

Plasma protein adsorption on the particle surface was studied by incubation of particles in rat blood plasma at 37C for 2.5 h followed by particle separation (centrifugation in glycerol gradient) and investigation by means of electrophoresis and blotting [6].

The particle (MION-46) surface was altered by poy-L-lysine (molecular weight = 10 000 Dalton) immobilization via the periodate / borohydride method [7] and, then, by complete succinilation of poly – L – lysine aminogroups. The amount of poly-L-lysine molecules on the particle surface was controlled by trinitrobenzensulfonate (TNBS) reaction [8].

Dextran-grafted poly-L-lysine polymers were prepared by conjugation of periodate-oxidized dextran (molecular weight 10 000 Dalton, 20 % units oxidation) with DTPA-labelle poly-L-lysine (molecular weight 10 000 – 70 000 Dalton) followed by borohydryde reduction. Ther polymers were labeled with ^{111}In by ligand exchange in 0.1 M citrate solution, pH = 5.3.

The *in vivo* mass transfer processes of radiolabeled particles and polymers after intravenous administration were examined in rats by dynamic γ-scintigraphy as described in article [4]. The final biodistributions were investigated following conventional techniques after stabilization of radioactivity levels in tissues.

Results and Discussion

Particles size and structure. Both preparations, MION-37 and MION-46, were found to consist of particles comprising, mainly, one electron dense core. The diameter of core was 4.4 to 6 nm for both preparations, while the total diameter in water (by LLS) was 21 ± 6 for MION-46 and 22 ± 6 for MION-37.

The diameter of MION-46 containing reverse micelles was (29 ± 9) nm at W / AOT / O = 1 / 0.5 / 50, which is in good agreement with particle diameter in water (assuming that AOT comprises 2 nm of the diameter). Dehydration of particle-containing micelles leads to the decrease of micelle overall size to (11 ± 4) nm at W / AOT / O = 1 / 5 / 50. Since amounts of iron oxide and dextran in MION-46 were found to be almost equal (iron oxide / dextran = 39 / 61 w / w), these data are consistent with hypothetical structure of the particle as a single iron oxide core surrounded by hydrated layer of dextran molecules in «wet» micelles and by almost dehydrated collapsed dextran layer in «dry» micelles. It is notable that under the same conditions (suspension in octan / AOT), the beads of cross-linked dextran gel (Sephadex G-200) almost do not change their size when the concentration of AOT is increased. This fact allows us to conclude that the dextran layer at the particle surface is more flexible than Sephadex G-200 gel and the structure of dextran layer corresponds, most likely, to a brush-like system of dextran molecules attached to the central core.

Particle interaction with blood protein. Despite similar total size core diameter and dextran content, particles of MION-37 and MION-46 have shown different behavior in relation to blood proteins. While MION-37 apparently adsorb almost all plasma proteins in amounts proportional to their concentration in plasma (including transferrin and immunoglobulines). MION-46 binds preferably the 3rd component of complement (C3). The total amount of protein adsorbed from plasma was 0,5 mg / mg iron for MION-37: this difference is even more pronounced in dilute plasma. The data obtained suggest that protein is bound to MION-37 and MION-46 particles via different mechanisms. While adsorption of a diversity of proteins on MION-37 appears to be the result of active nonspecific protein binding (probably initiated by the exposed iron hydroxide moieties), selective accumulation of C3 at MION-46 particles indicates better protection of the superparamagnetic core by dextran molecules, and binding of C3 to these molecules via the reaction of OH-groups of dextran with thioster group of C3 [9].

Modeling particle surface structure. To test the hypothesis of a high degree of particle core protection by dextran molecules in MION-46, a model polymer was synthesized. Dextran molecules were densely grafted onto poly-L-lysine molecule, on order to obtain a structure simulating the flexible, randomly organized dextran layer hypothesized in section 3.1: the total size of polymer molecule was similar to that of MION-46 particles. Since the particle surface is believed to be responsible for its in vivo behavior, biokinetics and biodistribution of dextran-grafted polymer were expected to correspond to that of iron oxide particle perfectly covered by dextran molecules. Inner poly-L-lysine molecule of the model polymer was labelled with DTPA to provide convenient binding site for ^{111}In.

The biokinetics and biodistribution studies in rats and rabbits have shown that blood clearance rates of MION-46 and the model polymer were similar; the half-clearance time for both preparations varied from about 1 h at low administered dose (2 µg / kg body weight for polymer) to 3.1 h at high dose (100 mg / kg). The blood clearance curve of the polymer was practically monoexponential within the dose range from 0.2 to 20 mg / kg.

The biodistribution of the model polymer and MION-46 particles (performed 24 h after intravenous administration) was found to be almost the same. Unexpectedly, the highest accumulation of both preparations was found in lymph nodes. Mesenteric and para-aortic group of lymph nodes have shown average accumulation of 30 – 50 % dose / g tissue (more than 110 % dose / g in some nodes), popliteal and axillary groups accumulated 10 – 20 % dose / g; cervical, thoracic and other nodes accumulated between 2 and 20 % dose / g. The standard deviation of accumulatio0n was high (20 – 95 % of mean value for different groups). Only 1.5 – 2 % of dose / g tissue was found in the liver and 15 – 20 % of dose / g tissue in the spleen. Other tissue accumulated less then 0.2 % dose / g tissue.

The results of this experiment allow us to conclude that: (1) the biological properties of dextran layer of MION-46 particles are similar to those of the random dense structure formed by dextran in the model polymer molecule; (2) in view of their similar biological properties, these two structure are most probably, similar: (3) the iron oxide core of MION-46 does not (or minimally) influence particle biodistribution and, most likely, the particle core is protected by dextran molecules against direct interaction with plasma proteins and cells.

The unique biological properties of dextran-grafted polymers discovered in these experiments, make the in potentially useful as lymphotropic carriers of diagnostic agents and therapeutic drugs [10, 11].

Altering particle surface. To study the effect of particles surface modification on biokinetics and biodistribution MION-46 by means of poly-L-lysine immobilization on the particle surface (MION-PL) and subsequent complete succiniliation of this derivative (MION-PL-SU). Diffusional diameters of both preparations were 26 nm (by LLS) with narrow size distibution. The average number of poly-L-lysine molecules was 11 per particles, assuming poly-L-lysine molecular weight = 10 kDalton. While basis MION –46 particle were slightly negatively charged. MION-PL particles are positively charged (having more than 500 aminogroups per particle) and MION-PL-SU particle carry the same number of negatively charged carboxylic groups.

The experimental data clearly demonstrate the dependence of particle blood clearance rate and final biodistibution on particle surface character. Immobilization of polylysine molecules on the particle surface decreases half-clearance time to 1 – 2 min. The final biodistribution of the particles in also affected dramatically by the manipulations described: about 50 % of the injected dose of MION-PL was found in the liver, a significant amount in the spleen and almost no material was found in the lymph nodes. However, complete succinilation of MION-PL recovered the circulation time almost to the initial level (half-clearance time 40 – 50 min) and provided accumulation of the preparation in the lymph nodes.

The above data enable as to conclude that: (1) negatively charged and essentially neutral particle surface provides for the longest circulation time; (2) immobilization of macromolecules on the particle surface does not necessarily prevent their accumulation in the lymph nodes.

Conclusion

In summary, we conclude that colloidal particle and polymer molecule of the same size and similar outer structure possess similar biokinetics and biodistribution due to their similar interactions with biological objects. Therefore, soluble polymers can be used in biological models as model compounds of colloidal particles and vice versa.

Particle surface structure, as a biodistribution-defining factor, may play a significant role in the development of the superparamagnetic MRI contrast agents. Since superparamagnetic particles are often used as basic compounds for targeted contrast agents, their interactions in vivo should be necessarily taken into account. Long circulation of a «targeted» dextran-coated particles may allow the exhibition of their transfer not only to the expected target site, but also to the lymphatic tissues. On the other hand, dextran-grafted particles may be effectively used in model experiments: as soon as the biodistibution of the basic preparation is known, the influence of the transporting molecules attached to the particle surface on the biokinetics and biodistribution may be easily evaluated in comparative studies.

References

[1] R. Weissleder, M. Papisov. // *Rev. Magn. Res. Medicine*. 4 (1992) 1.

[2] R. Weissleder, A. Bogdanov, M. Papisov. // *Magn. Res. Quart*. 8 (1992) 55.

[3] R. Molday, D. Mackenzie. // *J. Immunol. Meth*. 52 (1982) 353.

[4] M. Papisov, V. Savelyev, V. Sergienko, V. Torchilin. // *J. Pharm*. 40 (1987) 201.

[5] M. Dubois, K. Gilles, J. Hamilton, P. Rehers, F. Smith. // *Anal. Chem*. 28 (1956) 350.

[6] Bogdanov, M. Papisov, R.Weisleder, T. Shen, T. Brady in *Abstr. Soc.* Magn. *Res. Med. 11th Ann Meeting* (Berlin, 1992) P. 864.

[7] Dutton, K. Totuyasu, S. Singer. // *Proc. Nat. Acad. Sci*. 76 (1979) 3564.

[8] G. Means, W. Cangdom, M. Bender. // *Biochem*. 11 (1972) 3564.

[9] M. Panghurn. // *J. Immunol.* 142 (1989) 2759.

[10] M. Papisov, A. Bogdanov, R. Weissleder, E. Tsai, B. Schaffer, N. Molea, A. Bogdanova, H. Strauss, T. Brady in *Abstr. Soc. Magn. Res. Med. 11th Ann. Meeting* (Berlin, 1992), P. 567.

[11] N. Molea, R. Callahan, R. Wilkinson, E. Tsai, T. Brady, H. Strauss, M. Papisov. // *Eur. Nucl. Med.* 19 (1992) 590.

In: Preparation and Properties of Monomers, Polymers... ISBN 978-1-60021-557-5
Eds: A. Ballada and G. E. Zaikov, pp. 75-81

Chapter 8

INVESTIGATION OF MICELLIZATION AT NONIONIC SURFACTANTS IN THEIR SOLUTIONS

F. S. Bayburtskiy*[*,1], A. D. Korenev[2] and Yu. A. Novozhilov[2]

[1]N. M. Emanuel Institute of Biochemical Physics, Russian Academy of Sciences
4, Kosygin str., Moscow 119991, Russian Federation
[2]M. V. Lomonosov Moscow State University, Department of Chemistry,
Vorobyoby gory, Moscow, 119899 Russian Federation

ABSTRACT

Formation of micelles (micellization) in aqueous solutions of polyoxyethylene p-cumyl phenyl ether (PCPh (EO) $_{8-20}$) has been studied, from the surface tension data a function of concentration and temperature. Experimental data have been discussed according to change in enthalpy and entropy of micellization with temperature and polyoxyethylene chain length of surfactants.

Keywords: nonionic surfactants, critical micelle concentration (CMC), micellization.

INTRODUCTION

Unlike the nonionic surfactants of the type polyethoxylated alkylphenols with 8 and 9 carbon atoms in the alkyl chain which were widely investigated, the polyoxyethylene alkylphenyl ethers with a short alkyl chain have been scarcely studied [1].

As the number of ethylene oxide moles contained in a molecule has a great effect on the properties of these compounds, it has been considered proper to undertake a study of this category of surfactants as a function of the ethylene oxide chain length (EO).

Knowing the variation of the critical micelle concentration (CMC), surface properties and thermodynamic parameters at the micellization of surfactants, one can draw some conclusions

---

* E-mail: Chembio@sky.chph.ras.ru, Bayburt@mail.ru

regarding the practical application of these compounds as well as their industrial development in the future.

In this paper an investigation of polyethoxylated p-cumylphenol with 8 – 20 ethylene oxide units in molecule is presented. The surface tension was measured as a function of concentration and temperature aver the 20 – 40°C temperature range. The surface tension values thus obtained were then used to discuss the variation of critical micelle concentration as a function of ethylene oxide chain and temperature, the surface structure and the energetics of micellization.

MATERIALS AND METHODS

The studied nonionic surfactants, ethoxylated p-cumylphenol (PCPh (EO) n) where n = 8–20 oxyethylene units were technical products. They were characterized by determining the oxydril index and were employed as they were, without any previous purification. The water used in the preparation of the solutions was distilled water.

A Du Nouy tensiometer was used to measure the surface tension of the aqueous surfactant solutions. Measurements were carried out between 20 and 40°C at 5°C increments. The constancy of temperature was better than ± 0.1°. The CMC values were determined from the sharp breaks in the surface tension vs. logarithm of surfactant concentration plots and expressed in mole / l.

RESULTS AND DISCUSSIONS

1. Critical Micelle Concentration

No matter the ethylene oxide chain length, all aqueous solutions of the studied compounds have typical curves for the surface tension vs. concentrations plots. The surface tension of (PCPh (EO) n) is characterized by a sharp decrease at low surfactant concentrations which is then followed by a constant value or, in the case of the long ethylene oxide chain compounds, a slight increase.

The determined CMC values are in the same range as those found for other ethoxylated alkylphenols [2], that is of the magnitude of 10^{-4} mole / l. It is known that for the most nonionic surfactants, the CMC increases when the molecule becomes more hydrophilic by increasing the ethylene oxide chain length [3]. In the case of the surfactants studied in the present paper, the CMC increases only up to 15 EO units, value of which it starts decreasing with a further increase of the hydrophilic part of the molecule (figure 1).

Taking into account that the p-cumylphenol has a very short alkyl chain of only 3 carbon atoms, this behavior is not surprising. It has already been noted with the nonionic surfactants having a short or intermediary hydrocarbon chain [4, 5]. The decrease of CMC with increasing the number of EO units in the molecule has also been pointed out at studying the acetate-type nonionic surfactants [6] and different ethers of polyethylene glycol's [7].

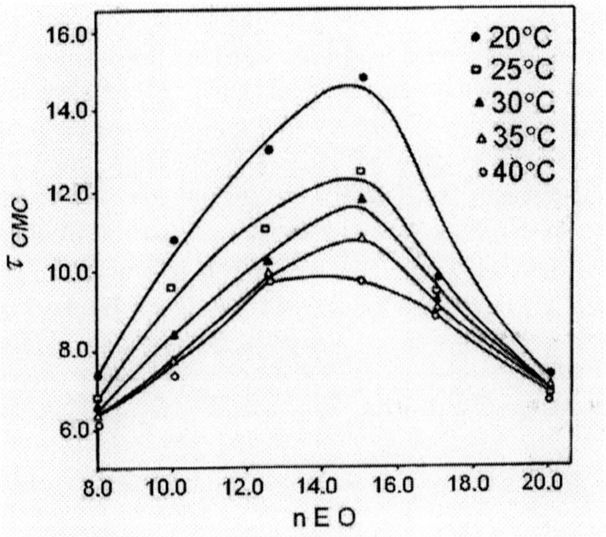

Figure 1. Variation of parameter τ_{CMC} depending on temperature.

It is commonly known that the hydrophilic part of the molecule, hydrated oxyethylene chain, hinders the aggregation as micelles, aggregation which is enhanced by the hydrophobic part of the molecules. It is normal for the CMC to increase with increasing the hydrophilic character of the molecule which is a result of the lengthening of oxyethylene chain. In other words, the «hydrophobic binding» loses in its importance when the number of ethylene oxide moles in p-cumylphenol increases from 8 to 15. The «hydrophobic binding» becomes important at a further increase of the oxyethylene chain, probably because of lower hydration caused by a chain coiling.

In figure 1 one can see that the CMC values for all compounds decrease with increasing the temperature and the peak becomes flat. This behavior confirms the importance3 of the oxyethylene chain hydration, which decreases with increasing the temperatures. As a consequence of decreasing the molecule hydrophilic the concentration of micelle formation decreases. Unlike the temperature effect on the aggregation of ionic surfactants (when a variation of the dielectric constant of solvent occurs), the aggregation in our case is governed by the «hydrophobic effect»: the transfer of hydrocarbon group from water to the nonpolar media, is normally an endothermic process, the hydrophobic bindings, and then the micelles, become more stable as the temperature increases [8].

It is also ascertained that, the more hydrophilie the molecule (higher CMC), the greater the differences between the CMC's at different temperatures. At 20 moles of ethylene oxide, chain lengthat which the hydration effect we presume taht is reduced due to a chain coiling, the temperature effect on the CMC is minimum.

2. Surface Properties

The surface tension of CMC (τ_{CMC}) decreases linearly with temperature, no matter the ethylene oxide chain length (figure 2). This is a normal behavior as much as nonionic surfactants we concerned, because of their surface activity which increases with temperature.

The same trend is common to the soap solution of saturated carboxilated acids [9]. The linear decrease of τ_{CMC} with temperature once again takes into consideration the importance of the «hydrophobic effect», that is the water orientation around the hydrophobic part of molecul or «iceberg» formation [10].

The water structure breaks down with increasing the temperature when dealing with both ionic and nonionic surfactants. The hydrocarbon chain becomes more mobile, but, in the same time, the ethylene oxide chain dehydration increases the hydrophobic character of the molecules, like the hydrocarbone chain increasing [11]. The lowest τ_{CMC} was obtained for the compound having 17 ethylene oxide moles, and the highest for the 12.5 moles – containing one. The surface tension data were then used to calculate the surface excess, Γ, area per molecule, A, of ethers and area per EO unit, A_{EO}, in the surface adsorption layer, assuming the latter to be monomolecular and log [a] = log [c]. Gibbs simple adsorption equation was employed:

$$-\Gamma = \frac{1}{2.303RT} * \frac{d\tau}{d\log[c]}.$$

Area per molecule was calculated according to the relation: A = 1 / N_A Γ, where N_A is Avogadro's number.

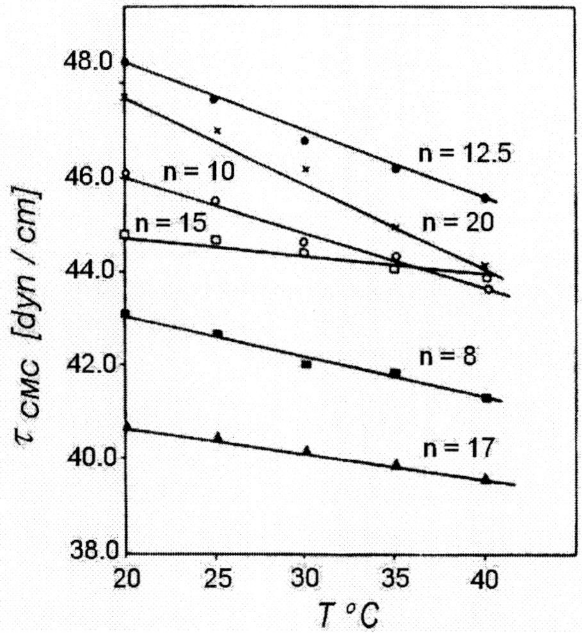

Figure 2. CMC variation depending on size polyoxyethylene lengths of a circuit (n EO).

It can be seen from the Table 1 that area per molecule at all temperatures increases with increasing the ethylene oxide chain length, behavior similar to that found out in the case of other non-ionic polyethoxylated compounds. The area per molecule is larger than the cross-out one of the hydrocarbon chain (20.5 $Å^2$), or the alkylated benzene ring (39.5 $Å^2$). That means that the ethylene oxide chain determines the area per molecule by its orientation in a position parallel to the surface.

Table 1. The surface and micellar properties of PCPh (EO) n PCPh (EO) $_8$

T°C	Γ (moles / cm ^{2} * 10 ^{10})	A (Å ^{2})	A $_{EO}$ (Å ^{2})	CMC (moles / l * 10 ^{4})
20	2.46	67.7	8.5	7.3
25	2.90	57.4	7.2	6.7
30	3.03	55.0	6.9	6.5
35	3.09	53.9	6.7	6.3
40	3.03	54.8	6.8	6.2

PCPh (EO) $_{10}$

T°C	Γ (moles / cm ^{2} * 10 ^{10})	A (Å ^{2})	A $_{EO}$ (Å ^{2})	CMC (moles / l * 10 ^{4})
20	2.26	73.7	7.4	10.8
25	2.30	72.5	7.2	9.6
30	2.74	60.8	6.1	8.4
35	2.83	58.6	5.9	7.7
40	2.71	61.5	6.2	7.3

PCPh (EO) $_{12,5}$

T°C	Γ (moles / cm ^{2} * 10 ^{10})	A (Å ^{2})	A $_{EO}$ (Å ^{2})	CMC (moles / l * 10 ^{4})
20	2.23	74.7	5.9	10.7
25	2.29	72.8	5.8	11.0
30	2.13	78.3	6.3	10.1
35	2.16	76.7	6.1	9.9
40	2.27	73.4	5.8	9.8

PCPh (EO) $_{15}$

T°C	Γ (moles / cm ^{2} * 10 ^{10})	A (Å ^{2})	A $_{EO}$ (Å ^{2})	CMC (moles / l * 10 ^{4})
20	2.10	79.4	5.2	14.8
25	2.27	73.4	4.9	12.5
30	1.94	85.9	5.7	11.8
35	1.88	88.3	5.8	10.8
40	2.02	82.5	5.5	9.7

PCPh (EO) $_{17}$

T°C	Γ (moles / cm ^{2} * 10 ^{10})	A (Å ^{2})	A $_{EO}$ (Å ^{2})	CMC (moles / l * 10 ^{4})
20	1.60	104.2	6.1	9.8
25	1.59	104.8	6.2	9.4
30	1.38	120.7	7.1	9.0
35	1.64	101.2	5.9	8.9
40	1.55	107.5	6.3	8.8

PCPh (EO) $_{20}$

T°C	Γ (moles / cm ^{2} * 10 ^{10})	A (Å ^{2})	A $_{EO}$ (Å ^{2})	CMC (moles / l * 10 ^{4})
20	1.56	106.8	5.3	7.9
25	1.58	105.4	5.3	7.3
30	1.23	135.5	6.8	7.2
35	1.31	127.2	6.3	7.0
40	1.28	130.0	6.5	6.8

The surface occupied by an EO unit decreases with increasing the ethylene oxide chain length up to about 17 EO moles, value beyond which it starts increasing. This peculiar variation can be explained by a lower degree of adsorption of the ethylene oxide end on the surface, which is more and more attracted by the water as the hydrophilic of molecule increases.

Over the 15 EO moles, according to a mechanism similar to that pointed out the case of the CMC variation, the ethylene oxide chain can be adsorbed in a greater extent on the surface, one EO monomer occupying a larger area.

3. Thermodynamic Properties

Two general theories have been proposed to account for the thermodynamics of micellization: the mass action model and the phase separation model. In principle, the two models are identical. However, if the aggregation number is small, the mass action model is used, while if the aggregation number is large, the phase separation model is applied. Since the aggregation numbers of the micelles of ethoxylated alkyl phenols are large [12, 13], the phase separation model has been applied to the present results. Consequently, a Clausius-Clapeyron equation has been derived for the determination of the heat of micelle formation (ΔH_M) from the temperature dependence of the CMC:

$$-\frac{\Delta H_M}{RT^2} = \left(\frac{\partial \ln CMC}{\partial T}\right)P$$

Since the micellization process is considered a phase transition, $\Delta G_M = 0$ and thus: $\Delta S_M = \Delta H_M / T$. The enthalpies, ΔH_M, and entropies, ΔS_M, of micelle formation, within the studied temperature range, are presented in Table 2.

One can see that the ΔH_M values are positive and they steadily decrease with increasing the temperature. This is a general trend for all studied ethoxylated compounds and for other ethoxylated of alkyl phenols as well [14, 15]. The entropy variation is similar to that of the heat of micellization.

Table 2. Thermodynamic parameters. ΔH_M (kcals / mole) / SM (cals / degree · mole)

Formula	20°C	25°C	30°C	35°C	40°C
PCPh $(EO)_8$	4.8 / 16.4	2.1 / 6.9	1.4 / 4.6	0.7 / 2.2	0.5 / 1.5
PCPh $(EO)_{10}$	8.7 / 29.8	5.5 / 18.5	3.7 / 12.1	2.3 / 7.5	1.8 / 5.9
PCPh $(EO)_{12.5}$	3.2 / 10.9	1.4 / 4.6	0.7 / 2.2	0.5 / 1.5	0 / 0
PCPh $(EO)_{15}$	7.1 / 24.3	5.1 / 17.0	2.8 / 9.1	1.6 / 5.2	1.2 / 3.7
PCPh $(EO)_{17}$	3.7 / 12.5	2.1 / 6.9	0.9 / 3	0.5 / 1.5	0 / 0
PCPh $(EO)_{20}$	3.0 / 10.2	1.8 / 6.1	0.9 / 3	0.7 / 2.2	0.2 / 0.7

It is generally known that there are two main processes which oppose one another during the micellization [10] the destruction of the «iceberg» water structure about the monomer units of surfactant during their incorporation into the micelle, and the arrangement of randomly oriented monomer units into a well-ordered micellar structure. From on entropy viewpoint, the former process leafs to a positive variation in entropy during the transition from a monomer to a micellar state, while the latter process leads to a negative variation in entropy. The value of entropy is then a function of the predominating process.

The decreasing in micellization entropy of all studied compounds leads to the conclusion that the destruction of the «iceberg» structure loses in importance with increasing the temperature.

The heat and entropy of micellization increase with the ethylene oxide chain length up to 12 – 15 EO units; beyond this value the two parameters trend to a constant value. The initial increase of the thermodynamic parameters can be explained by the increase in molecule hydrophilic, that is, the transfer of monomer from water towards the micelle needs a higher energy. The entropy increases in the same time as the destruction of «iceberg» structure becomes the most important process. The subsequent decrease in ΔH_M and ΔS_M, noticed in the case of other polyethoxylated alkyl phenols as well [15], confirms the CMC variation as a function of EO unit number. Due to coiling of the ethylene oxide chain, i.e. the decrease in molecule hydrophilic, the necessary micellization energy, is lower, the aggregation process becoming more important than that of destruction of the «iceberg» structure.

CONCLUSION

Formation of micelles (micellization) in aqueous solutions of polyoxyethylene p-cumyl phenyl ether ($PCPh(EO)_{8-20}$) has been studied, from the surface tension data a function of concentration and temperature. Experimental data have been discussed according to change in enthalpy and entropy of micellization with temperature and polyoxyethylene chain length of surfactants.

REFERENCES

[1] R. Heusch, N. Fette, P. Seifen // *Anstrichm.* 72 (11) (1970) 969.

[2] P. Becher. «Nonionic surfactants» (editor M. J. Schick), 1967, P. 478.

[3] H. Lange. «Nonionic surfactants» (editor M. J. Schick), 1967, P.443.

[4] P. Becher. // *J. Phys. Chem.* 63 (1959) 1675.

[5] J. Carless, R. Challis, B. Mulley. // *J. Colloid Sci.* 19 (1964) 201.

[6] T. Kuwamura, H. Takahashi. // *Bull. Chem. Soc. Japan.* 45 (2) (1972) 617.

[7] St. Kucharski. // *Journal of Polymers.* 250 (1972) 231.

[8] M. Enerson, A. Holtzer. // *J. Phys. Chem.* 71 (1967) 322.

[9] N. Kostova, Z. Markina, P. Rebinder, A. Kuzmina. // *Kolloid Zh.* 30 (1968) 249.

[10] H. Frank, M. Evans. // *J. Chem. Phys.* 13 (1945) 507.

[11] K. Tamaki.// *Journal of Polymers.* 252 (1974) 547.

[12] D. Hall, B. Pethica. «Nonionic surfactants» (editor M. J. Schick), 1967, P. 516.

[13] J. Corkill, J. Goodman, P. Harrold. // *Trans. Faraday Soc.* 60 (1964) 202.

[14] E. Crook, G. Trebbi, D. Fordyce. // *J. Phys. Chem.* 68 (1964) 3592.

[15] Ray, G. Nemethy. // *J. Phys. Chem.* 75 (1971) 809.

In: Preparation and Properties of Monomers, Polymers… ISBN 978-1-60021-557-5
Eds: A. Ballada and G. E. Zaikov, pp. 83-88

Chapter 9

ASSOCIATION OF MOLECULES AND FORMATION OF MICELLES IN SOLUTIONS IONIC SURFACTANTS

***F. S. Bayburtskiy**[*,1], **A. D. Korenev**[2] **and Yu. A. Novozhilov**[2]*
[1]N. M. Emanuel Institute of Biochemical Physics, Russian Academy of Sciences
4, Kosygin str., Moscow 119991, Russian Federation
[2]M. V. Lomonosov Moscow State University, Department of Chemistry,
Vorobyoby gory, Moscow, 119899 Russian Federation

ABSTRACT

The results of physical and chemical investigations of the surfactants show complex phenomena: the formation of colloid associates or submicellar species, below the critical micellar concentration and the formation of micelles or aggregates over the critical micellar concentration.

Key words: surfactant, colloid associates, submicellar species, critical micellar concentration.

INTRODUCTION

General properties of surfactant solutions are reviewed by Preston [1]. Some correlating principle of detergent action show the existence of a critical washing concentration; a value characteristic for each surfactant. In patent literature detergent formulations appear abundantly, but it is well known that many of the representative samples contain surfactant in amounts which assure the formation of surfactant micelles in laundry machine or manual washing [2]. Results of investigation of the «surfactant / surfactant» systems, below critical micellar concentrations, indicate as a reasonable possibility the appearance of submicellar structures [3 – 7]. In such case the critical concentration for micelle formation corresponds to the agglomeration concentration at which the surfactant associates lose specific and

* E-mail: Chembio@sky.chph.ras.ru, Bayburt@mail.ru

characteristic colloids properties. Within scope of this problem the intent of this work is to provide an updated scheme of the colloid characteristics of surfactant solutions below the critical micellar concentration in order to better understand the physical and chemical mechanisms, as well as surfactant mixture processes in aqueous media.

Materials and Methods

The following systems were analyzed: 1) sodium sulfate (0.014 – 30 %) in aqueous media, in calcium chloride (10 %), in sodium tripolyphosphate (2 g NaTPP / l) and in solution containing $CaCl_2$ (10°d) + (2 g / l) NaTPP at 40°C; 2) cationic surfactant n-dodecylammonium nitrate $DDANO_3$ + anionic surfactant sodium n-dodecylsulfate $NaDDSO_4$ in aqueous media; 3) $DDANO_3$ in aqueous media; 4) $NaDDSO_4$ + electrolyte in aqueous media and 5) sodium n-dodecylbenzensulfonate $NaDDBSO_3$ in aqueous media containing various electrolyte. All systems were prepared by direct mixing of equal volume of fresh surfactant solutions. Prepared systems were aged at constant temperature of 20 – 40°C for 10 minutes prior to the analysis. The surface tension determinations were conducted on an interfacial tensiometer. Results are presented as the average value of 7 measurements. Tyndallometric values were registered by photometer with a turbidimetric extension. The electrophoretic mobility and particle charge were determined by the aid of an ultramicroscope with a double capillary electrophoretic cell: the mean of seven samples was presented as the results. Equivalent conductivity was determined by means of a high frequency conductometer (used conductivity cells with a platinum electrodes were completely in a thermostated body).

Results and Discussion

An attempt was made to elucidate conditions necessary for the formation of submicellar species of surfactants with colloid properties. In aqueous media, below the critical concentration for micelle formation a series of experiments were performed: light scattering, electrophoretic mobility, conductivity and interfacial tension measurements. The interfacial tension σ of the «zero» systems, i.e. of Na_2SO_4 in aqueous solutions at 40°C shows an independence on the Na_2SO_4 concentration over 10 %, in the presence $CaCl_2$ (10°d = 3,57 mole-equivalent / liter) the constant σ value over 5 % of Na_2SO_4, independence on Na_2SO_4 for the mixed solution $CaCl_2$ (10°d) + (2 g / l) NaTPP + (0,014 – 30 %) Na_2SO_4 and for systems with NaTPP (2 g / l) a constant decrease of σ value with an increase in Na_2SO_4 concentration. A constant σ value for the $CaCl_2$ + NaTPP solutions of various Na_2SO_4 concentrations indicates that the interfacial changes caused by the change of surfactant influence. Pure sodium dodecylsulfate $NaDDSO_4$ solutions at high sodium sulfate concentration platted versus Na_2SO_4 concentrations (figure 1 – dashed line) show linear course for the critical micellar concentration. The same slopes were obtained for the *b*, *c* and *d* conditions (see figure 1).

Commercial sodium n-dodecylbenzensulfonate $NaDDBSO_3$ solutions depending on the chemical conditions in the liquid phase shows complex dependence of critical micellar concentration on the sodium sulfate concentration (figure 1 – *a, b, c, d*). Since σ value varies differently for *a, b, c* and *d* system and because in the system without $NaDDSO_4$ no equivalent changes are registered, it is reasonable to assume that the main observed phenomena depends on the changes of surfactant concentrations, but differently for the $NaDDSO_4$ and $NaDDBSO_3$. Results of interfacial tension measurements indicate that the surfactant species present in various concentrations are main factor causing observed phenomena. For this reason a minute analysis of interfacial tension as a function of $NaDDBSO_3$ concentration in 10 15 and 20 % Na_2SO_4 solutions was performed. The results (figure 2 – *a*) show five different slopes from A to E, all for below critical micellar concentration (indicated between A and C slopes and C and D slopes in figure 1). As the transition or critical surfactant concentration of associate formation.

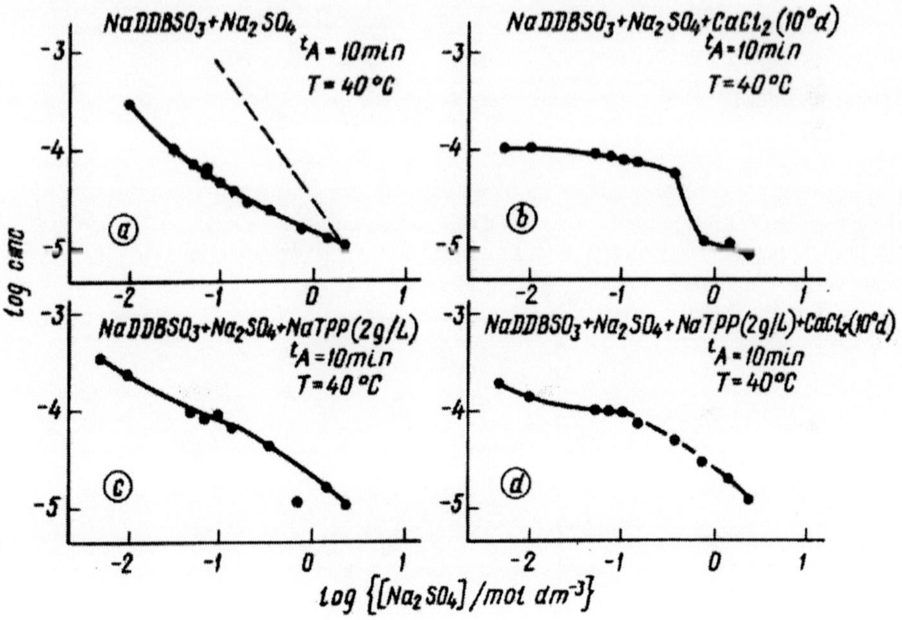

Figure 1. Systems: a) $NaDDBSO_3$+ Na_2SO_4; b) $NaDDBSO_3$ + Na_2SO_4 + $CaCl_2$; c) $NaDDBSO_3$ + Na_2SO_4 + NaTPP and d) $NaDDBSO_3$ + Na_2SO_4 + NaTPP + $CaCl_2$ all in aqueous media at 40°C. Critical micellar concentration (*CMC*) is plotted versus concentration of Na_2SO_4 both in log scale. Dashed line represents the schematic data for the systems with $NaDDSO_4$ instead of $NaDDBSO_3$.

Specially prepared and purified surfactant n-dodecylammonium nitrate $DDANO_3$ in aqueous media was analyzed as an oppositely charged cationic surfactant. Measured conductivity as a function of the surfactant concentration shows (figure 2 – b) six different slopes, also at surfactant concentrations for below critical micellar concentration. According to the Hiskey-Downey results the same cocnclusion can be drawn for associate formation [8]. Di,- tetra- and polymer associates op surfactant possess the new colloid characteristics which can be verified in the following manner. Placing both $DDANO_3$ (positive cationic associate z^+) and $NaDDSO_4$ (negative anionic associate z^-) or $NaDDBBSO_3$ into contact, colloid species will form by mutual interaction.

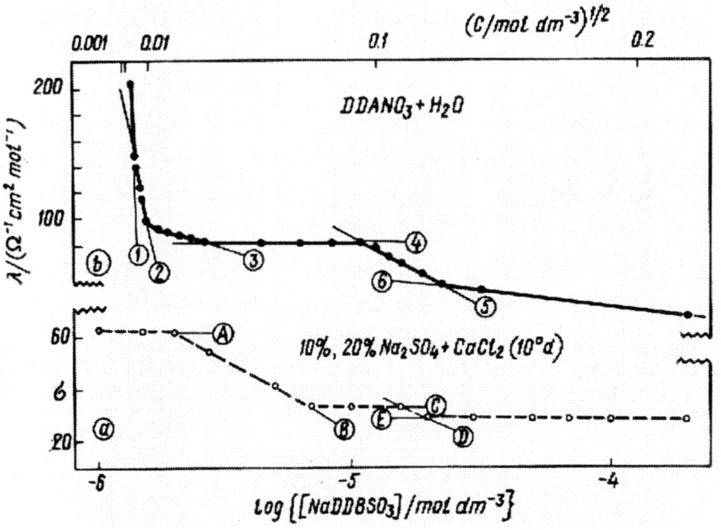

Figure 2. System 2 – a: 10 % or 20 % (per weight) Na_2SO_4 + (10°d) $CaCl_2$ + $NaDDBSO_3$. Surface tension σ - (dynes / cm) is plotted versus $NaDDBSO_3$ concentration (Log scale). System 2 – b: n-Dodecylammonium nitrate $DDANO_3$ in aqueous media. Equivalent conductivity $\pi / (\Omega^{-1}\ cm^2\ mole^{-1})$ is plotted versus square root of $DDANO_3$ concentration.

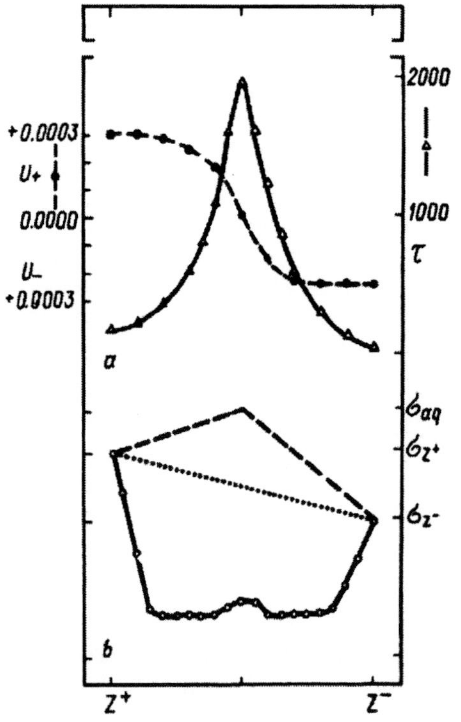

Figure 3. Systems: at constant total surfactant concentration of $CDDANOO_3$ (z^+) + $CNaDDSO_4$ (z^-) = 0,0001 M surface tension σ (open circles), tyndallometric value τ (triangles) and electrophoretic mobility $U_\pm$ (full cicrles) are plotted versus z^+, z^- part respectively. The samples differed 10 per cent in a combination of z^+ and z^- components.

Results presented in figure 3 show the formation of typical Tyndall maximum for turbidity measured as a function of the z^+ / z^- ratio (figure 3 – triangles).

According to the Tamamushi concept [9], the electrostatic forces between ionic groups involving van der Waals attractions between carbon tails of the cationic and anionic surfactant may be responsible for the observed phenomena. Our hypothesis of the formation of submicellar associates with distinguished colloid properties is supported by the electrophoretic mobility $U_{\pm}$ measured as a function of z^+ / z^- ratio shows a typical electrophoretic curve. The colloid aggregates formed by mutual interaction between cationic and anionic surfactant associates act very strongly on the interface layer causing a change of interface tension (figure 3 – open circles). The resulting curve confirms our approach, because for «physical mixture» z^+ and z^- without colloid associates the resulting interface tension must be calculated as an average σ value (figure 3 – dotted line); «chemical neutralization» between cationic and anionic surfactants, as a consequence gives the resulting interfacial tension curve with a maximum σ value corresponding to the σ value of water σ (H_2O) (figure 3 – dashed line); since neither «physical mixture» nor the «chemical neutralization» between cationic and anionic surfactants, as a consequence gives the resulting interfacial tension curve with a maximum σ value corresponding to the σ value of water σ (H_2O) (figure 3 – dashed line); since neither «physical mixture» nor the «chemical neutralization» was obtained, the resulting curve (open circles) indicates that the colloid interaction between surfactants of concentrations below critical micellar concentration appear as a consequence of the present cationic and anionic surfactant associates.

CONCLUSION

The results of physical and chemical investigations of the surfactants show complex phenomena; the formation of colloid associates or submicellar species, below the critical micellar concentration and the formation of micelles or aggregates over the critical micellar concentration.

REFERENCES

[1] W. C. Preston. // *J. Phys. Colloid. Chem.* 52 (1984) 84.

[2] W. G. Culter, R. C. Davis. *Detergency, Theory and Test Methods*. (editor Marcel Dekker), Inc. New York, 1972, Part I. P. 14 – 27.

[3] V. Tomasic, N. Filipovic-Vincekovic, B. Kojic-Prodic, N. Kallay. // *Colloid and Polymer Science,* 269 (1991) 1289.

[4] V. Tomasic, N. Filipovic-Vincekovic, D. Sktric. // *Ber Bunsengens Physical Chemistry*, 95 (1991) 1646.

[5] N. Filipovic-Vincekovic, M. Bujan, D. Dragevic, N. Nekic. // *Colloid and Polymer Science*, 273 (1995) 182.

[6] M. Bujan, N. Vdovic, N. Filipovic-Vincekovic. // *Colloid and Surface Science,* 118 (1996) 121.

[7] D. Dragevic, M. Bujan, Z. Grahek, N. Filipovic-Vincekovic. // *Colloid and Polymer Science*, 273 (1995) 967.

[8] C. F. Hiskey, T. A. Downey. // *J. Physical Chemistry*, 58 (1954) 835.

[9] B.Tamamushi, K. Tamaki. // *Colloid Journal*, 163 (1959) 122.

In: Preparation and Properties of Monomers, Polymers… ISBN 978-1-60021-557-5
Eds: A. Ballada and G. E. Zaikov, pp. 89-98

Chapter 10

THE INTERACTION OF SURFACTANTS WITH ION POLYMERIC SORBENTS

***F. S. Bayburtskiy*[*,1], *A. D. Korenev*[2] *and Yu. A. Novozhilov*[2]**

[1]N. M. Emanuel Institute of Biochemical Physics, Russian Academy of Sciences
4, Kosygin str., Moscow 119991, Russian Federation

[2]M. V. Lomonosov Moscow State University, Department of Chemistry,
Vorobyoby gory, Moscow, 119899 Russian Federation

ABSTRACT

The isotherms of equilibrium adsorption on weakly and strongly basic macroporous polymeric ion exchangers of sodium dodecylsulfate, sodium dodecylsulfonate and sodium di (n-octyl) – sulfosuccinate were obtained. It was established that the adsorption isotherms as well as the maximum adsorption capacity depends strongly on the adsorbent nature and hydrophilic-lipophilic balance (HLB) value of the surfactants. It was pointed out that for weakly basic ion exchangers a sharp upwards sloping of the adsorption curve occurs at a certain equilibrium concentration. At this particular which has been called «critical hemimicellar concentrations» of the systems, (CMHC)s, two-dimensional aggregates begin to form on the solid surfaces as a result of hydrophobic interactions between the non-polar parts of surfactant molecules. The (CMHC)s values decrease by increasing the polar character of the adsorbent surface. The maximum adsorption coefficient for the studied surfactants fall in the sequence: weakly basic ion exchangers in OH^- form < weakly basic ion exchangers in Cl^- form < strongly basic ion exchangers.

Key words: surfactants, critical hemimicellar concentration, adsorption, ion exchangers.

---

[*] E-mail: Chembio@sky.chph.ras.ru, Bayburt@mail.ru

INTRODUCTION

During the last decades, a new class of polymeric ion exchangers with macroporous or macrolattice structure [1 – 3] has been synthesized. The main difference between classical gel-type ion exchangers and macroporous ion exchangers consist in a very high porosity and a larger surface area of the last even in the dry state. Each particle of macroporous ion exchangers can be considered as being made up of an enormous number of microspheres in between which pores of thirty to several hundreds angstroms one accommodate, this structure having a specific surface area of up to 9×10^5 m^2 / kg.

The synthetic macroporous ion exchangers have proved to be successfully used for removal of organic substances mainly surfactants, from waste waters. Besides their higher capacity to retain organic materials, they can be easier and more complete regenerates than the gel-type ion exchangers.

The synthesis of macroporous ion exchangers consist in copolymerization of two vinyl monomers, one with two functional groups in order to create cross-links, in the presence of a component that is a good solvent for the monomer and poor for the polymer, followed by introduction of ionogenic groups by classical means [4, 5]. The properties of macroporous ion exchangers are determined by the porous structure of skeleton as well as the nature of iongenic group. The adsorption of organic compounds and their regeneration on macroporous ion exchangers is determined by the nature of interaction between the two components. These interactions can be studied by means of adsorption isotherms [6]. We can presume that in this case the interactions are of electrostatic nature leading to an ion exchange and due to possibility to establish hydrophobic bands between nonpolar parts of adsorbents (skeleton) and nonpolar part of organic matter. To these two kinds of interactions one should add the possibility of hydrophobic interaction among adsorbed organic molecules (cooperative adsorption) [7, 8].

In order to elucidate the nature of interactions implied in adsorption of anionic surfactant on cationic macroporous ion exchangers, the adsorption of sodium dodecylsulfate ($DDOSO_3Na$), sodium dodecylsulfonate ($DDOSO_2Na$) and sodium di (n-octyl) sulfosuccinate (Monoxol OT) on three types of ion exchangers, strongly basic (SB) and weakly basic (WB) in OH^- and Cl^- form were obtained.

MATERIALS AND METHODS

Materials

Sodium dodecylsulfate was prepared from lauric alcohols of 99.8 % purity by reaction with chlorosulfonic acid in chloroform media, followed by neutralization with sodium hydroxide. The product was several times recrystallized from hot ethylic alcohol and extracted with petroleum ether for removing unreacted lauric alcohol. In order to check the purity of product, the isotherms of surface tension were obtained and the absence of the minimum near the critical micellar concentration (CMC) was asserted to a high purity.

The macroporous ion exchangers were copolymers of styrene-divynilbenzene, those weakly basic (WB 1 – WB 4) having dialkyl-aminogroups and that strongly basic (SB 1)-ammonium groups.

Procedure

The adsorption isotherms of $DDOSO_3Na$ on weakly basic ion exchangers in OH^- form are given in the figures 1 and 2, and the characteristic values of these isotherms are listed in table 1.

Table 1. The adsorption values of sodium dodecylsulfate on weakly basic anionites in OH^- form

Adsorbents	Adsorption coefficient at first plateau, a x 10^3 kmole / kg	Maximum adsorption, a x 10^3 kmole / kg	(CHMC)s x 10^3 kmole / kg
WB 1	0.32	1.00	5.4
WB 2	0.34	0.90	2.2
WB 3	0.12	0.82	2.2
WB 4	0.40	1.00	5.8

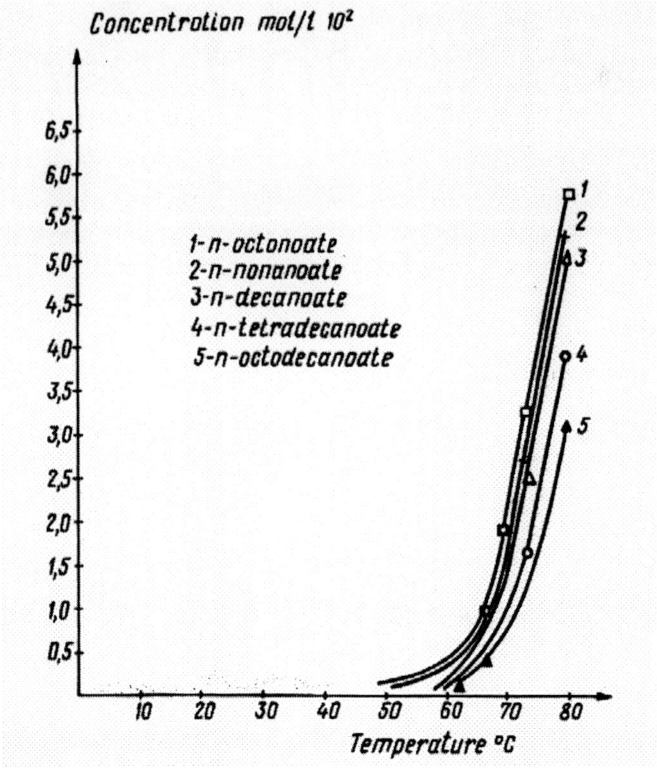

Figure 1. Adsorption isotherms of sodium dodecylsulfate on weakly basic anionites in OH^- format 298 K. (• – WB 1; ○ – WB 3).

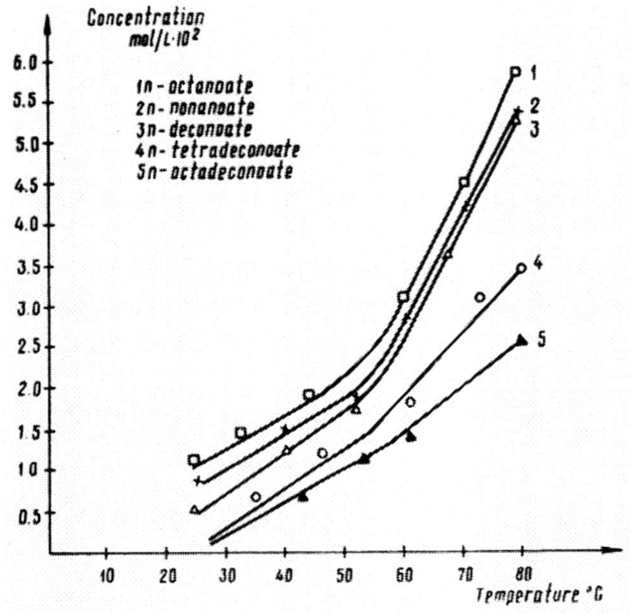

Figure 2. Adsorption isotherms of sodium dodecylsulfate on weakly basic anionites in OH^- form at 298 K. (• – WB 2; ○ – WB 4).

In figures 3 and 4 the adsorption isotherms of $DDOSO_2Na$ on weakly basic ion exchangers in Cl^- form are presented and the characteristic values are given in the Table 2. In order to compare the adsorption data, the adsorption isotherms of $DDOSO_2Na$ on Lewatite MP-62 and Amberlite Ura-93 are given in figure 5 and adsorption characteristic values tabulated in Table 2.

Table 2. The adsorption values of sodium dodecylsulfate on weakly basic anionites in OH^- form

Adsorbents	Adsorption coefficient at first plateau, *a* x 10 ^{3}kmole / kg	Maximum adsorption, *a* x 10 ^{3}kmole / kg	(CHMC)s x 10 ^{3} kmole / kg
WB 1	0.32	1.00	5.4
WB 2	0.34	0.90	2.2
WB 3	0.12	0.82	2.2
WB 4	0.40	1.00	5.8

Table 3. The adsorption values of sodium dodecylsulffate on weakly basic anionites in Cl^- form

Adsorbents	Adsorption coefficient at first plateau, *a* x 10 ^{3}kmole / kg	Maximum adsorption, *a* x 10 ^{3}kmole / kg	(CHMC)s x 10 ^{3} kmole / kg
WB 1	0.50	1.60	8.0
WB 2	0.65	1.80	7.0
WB 3	0.25	1.10	4.0
WB 4	0.50	1.25	5.2
IRA - 93	2.00	3.90	3.0

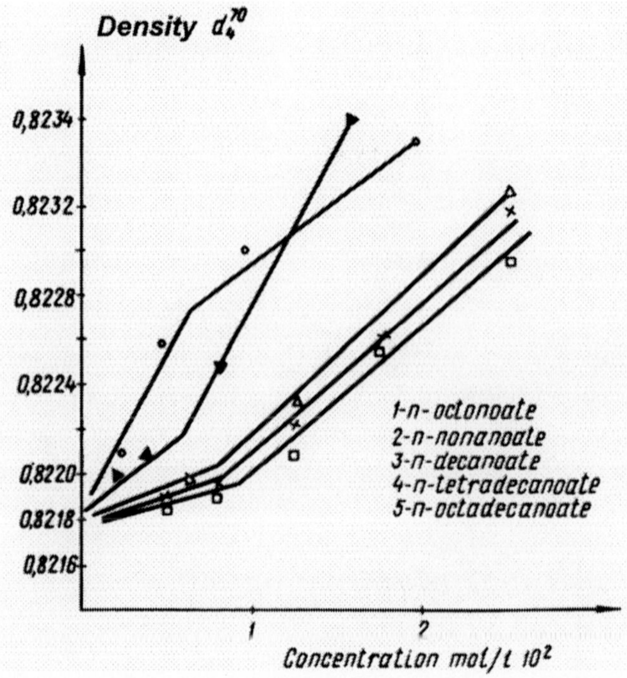

Figure 3. Adsorption isotherms of sodium dodecylsulfate on weakly basic anionites in Cl^- form at 298 K. (• – WB 1; ○ – WB 3).

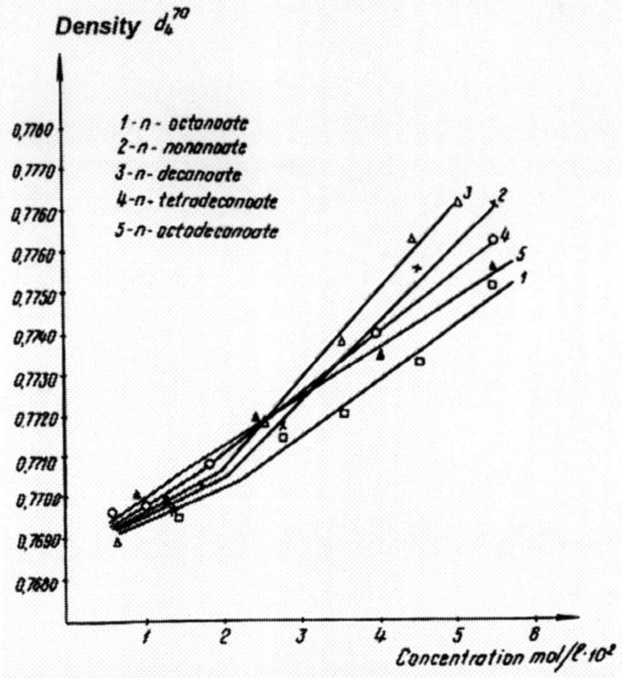

Figure 4. Adsorption isotherms of sodium dodecylsulfate on weakly basic anionites in Cl^- form at 298 K. (• – WB 2; ○ – WB 4).

Table 4. The hydrophobic character of the surfactants and maximum adsorption capacity

Compound adsorbed	Maximum adsorption, $a \times 10^3$, kmole / kg	
	on SB 1	on WB 2
Na_2SO_4	1.50	–
$C_{12}H_{25}OSO_3Na$	2.50	0.90
$C_{12}H_{25}OSO_2Na$	2.16	1.56
$C_{20}H_{37}O_4SO_3Na$	2.16	1.57

The adsorption isotherms of $DDOSO_2Na$ on strongly basic anionic exchangers (SB 1) together with exchange isotherm of Na_2SO_4 on the same adsorbent are given in figure 6. The adsorption isotherms of different anionic surfactants on weakly basic ion exchanger in Cl^- form are given in figure 7 and on strongly basic ion exchangers in figure 8. Table 3 contains the maximum adsorption capacity of the three studied surfactants on two types of adsorbents.

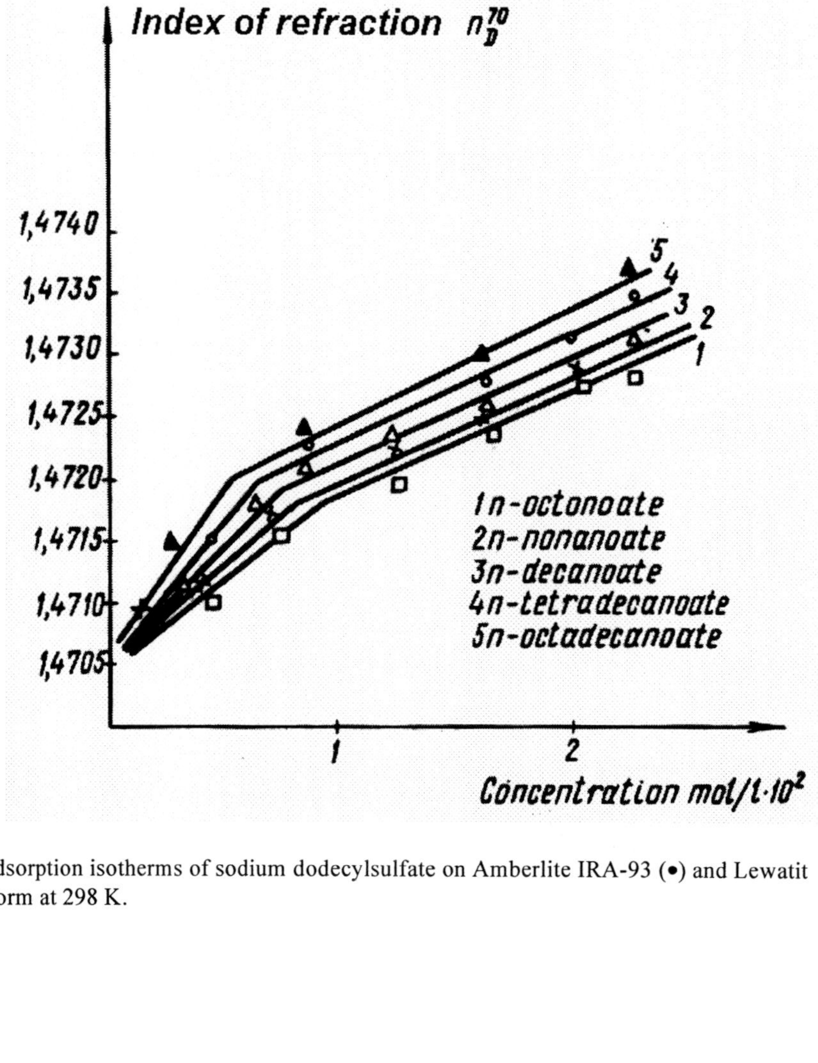

Figure 5. Adsorption isotherms of sodium dodecylsulfate on Amberlite IRA-93 (•) and Lewatit MP-62 (o) in Cl^- form at 298 K.

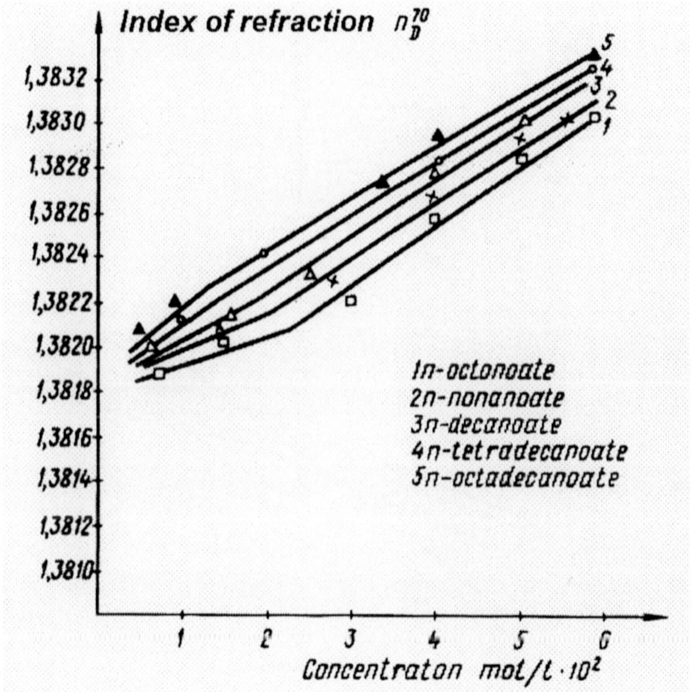

Figure 6. Adsorption isotherms of sodium dodecylsulfate (•) and sodium sulfate (○) on strongly basic anionite SB 1 at 298 K.

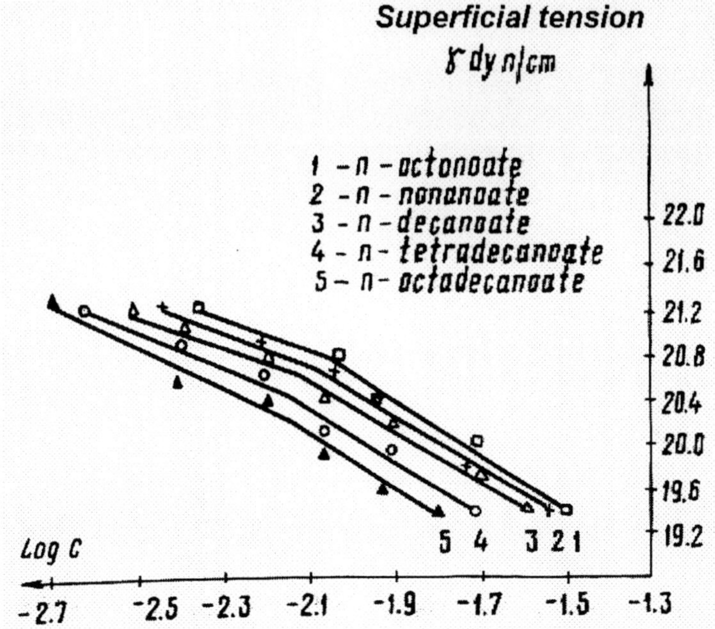

Figure 7. Adsorption isotherms on weakly basic WB 2 in Cl^- form at 298 K of: • – sodium dodecylsulfate; ○ – sodium dodecylsulfonate; ▲ – sodium dioctilsulfosuccinate; x – sodium dodecylsulfate on WB 2 in OH^- form.

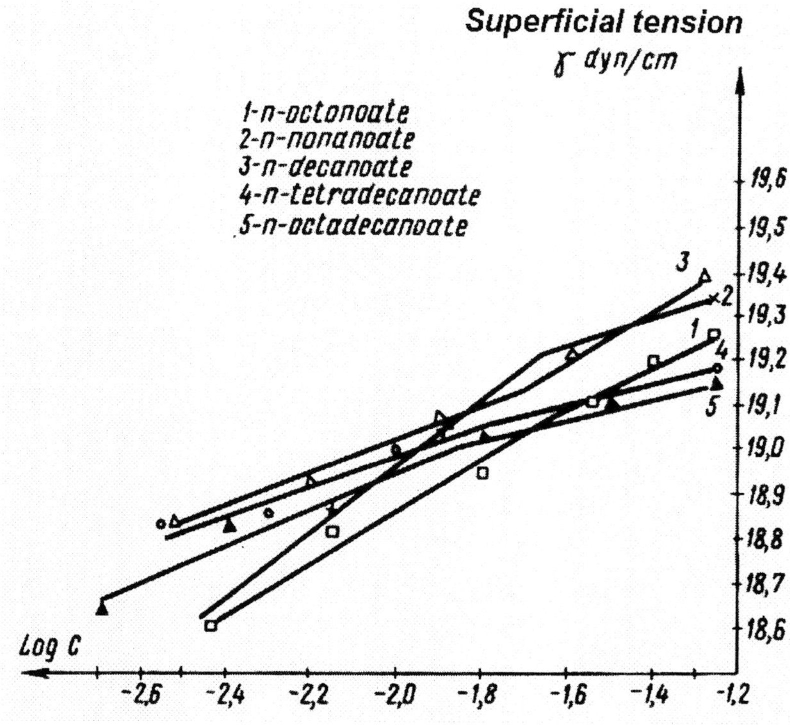

Figure 8. Adsorption isotherms on strongly basic SB 1 anionite at 298 K of: • – sodium dodecylsulfate; ○ – sodium dodecylsulfonate; Δ – sodium dioctilsulfosuccinate.

RESULTS AND DISCUSSION

1. Polarity of Macroporous Ion Exchangers

The form of adsorption isotherms of DDOSO3Na on weakly basic ion exchanger (WB) in OH$^-$ form (figure 1 and figure 2) is characteristic to L_4-type in Giles classification of adsorption isotherms of organic substance from solution [10]. The initial part of the isotherms shows that by increasing the coverage of the solid substrate with ions or molecules further adsorption on a vacant e cases that the adsorbed molecule lie on the place becomes more difficult. It is very likely in the surface. At higher concentrations, the adsorbed molecules of the paraffinic chain type can reorientate themselves be coming end-on, simultaneously with an adsorption increase due to the interactions between paraffinic chains (hydrophobic interactions) leading two-dimensional aggregates. According to Fuerstenau and coworkers [11, 12] the formation of two-dimensional aggregates is similar to micellization of amphilic molecules in aqueous solutions. These two-dimensional aggregates are called hemimicelles, and equilibrium concentration at which the adsorption coefficient increases sharply as a result of hydrophobic interactions between paraffinic chains of amphiphilic molecules on the surface is called the critical hemimicelle concentrations of the system (CMHC)s. The (CMHC)s is always lower than critical micellar concentration and depends on adsorbent.

Comparing the adsorption isotherms of $DDOSO_3Na$ on weakly basic ion exchangers in Cl$^-$ form (figures 3 and 4) with those of the same adsorbent in OH$^-$ form (figures 1 and 2) it can be pointed out the shape of the isotherms which remains unchanged. As for maximum adsorption value, this is higher for Cl^- form, as well as equilibrium concentration at which the sharp increasing of adsorption takes place (Table 1 and 2).

The adsorption isotherms of $DDOSO_3Na$ on weakly basic Lewatit MP-62 (figure 5) is different from all adsorption isotherms on weakly basic adsorbents, but is similar to the adsorption isotherms of other surfactants on weakly basic adsorbents. This isotherm is of L_2-type in Giles classification and in fact it represents the changes from L-type to H-type (high affinity).

The last type is characteristic to an adsorption on strongly basic ion exchangers.

Adsorption isotherms of $DDOSO_2Na$ on nonionic polymers [13] showed that the first part of those could not be obtained because the adsorption began only above a certain concentration of the surfactant, namely that where hydrophobic interactions between adsorbent and paraffinic chains became effective. These concentrations decrease by increasing the hydrophobic character of the polymeric adsorbent. In the case of povynilalcohol (a weakly hydrophobic polymer) the concentration of $DDOSO_2Na$ at which a sharp increase of adsorption occurs is 5.4 x 10^{-3} kmole / m^3, for methylcellulose 4.5 x 10^{-3} kmole / m^3 and for vinylalcohol-vinylacetate copolymer of 1.8 x 10^{-3} kmole / m^3 all under CMC of the surfactant (8.1 x 10^{-3} kmole / m^3 at 298K). A similar bahaviour was reported [14] for adsorption of cationic surfactants on «nonionic» polystyrene and «ionic» polystyrene obtained by emulsion polymerization using soaps as emulsifiers. For nonionic polystyrene, the first part of the isotherm is absent and for both types a sharp increase of adsorption is located under CMC of the surfactants.

The certain cases revealed of $DDOSO_2Na$ on strongly basic anionite (SB 1) is of H_1-type (figure 6) where practically the maximum of adsorption is reached at very low concentrations (from zero to 3 x 10^{-3} kmole / m^3). The H-type isotherms are characteristic to high affinity between adsorbent and adsorbate, so that even in dilute solutions the surfactant is completely adsorbed and the initial portion of the isotherm is vertical. This type of isotherms has been remarked for polymeric substances or big molecules as well as for amphiphilic molecules on different solid surfaces [18 – 20].

2. The Surfactants Nature

The affinity of anionic surfactant for weakly basic ion exchangers at low equilibrium concentrations is increasing in the order (figure 7): $DDOSO_3Na$ < $DDOSO_2Na$ < Manoxol OT. The maximum adsorption coefficients for the three surfactants are very close but they are reached at different equilibrium concentrations. These concentrations decrease in the above mentioned order. By increasing the surfactant affinity, the form of adsorption isotherms changes from L type ($DDOOSO_3Na$) to H type (Manoxol OT). The affinity increase corresponds to the increase in hydrophobic character of the surfactant [14, 20].

Maximum adsorption on strongly basic ion exchanger (figure 8) is reached at very low concentration due to the electrostatic interactions. Adsorption of $DDOSO_2Na$ and Manoxol OT gives a H_2-type while DDOSO3Na gives a H_1-type (continuous increasing). The maximum adsorption coefficients on strongly basic ion exchangers overpass constantly the ionic exchange capacity (Table 3).

Conclusions

The feature of the first part of adsorption isotherms is determined by the polarity of the solid surface. For weakly basic anionites, after reaching adsorption coefficient due to electrostatic interactions, a long plateau follow and then a sharp increase of adsorption as a result of hydrophobic interactions between the paraffinic chains of adsorbed molecules. There is a characteristic equilibrium concentration at which a sharp increase of adsorption begins. This value called (CMHC)S, depends on the surfactant nature as well as adsorbent and the two-dimensional aggregates resembling hemimicelles are supposed to form.

By increasing the hydrophobic character of the surfactant molecules the (CMBC)s decreases and the feature of adsorption isotherm changes that is the plateau tends to disappear. For all investigated polymeric ion exchangers the maximum adsorption capacity for surfactants overpasses the exchange capacity for Na_2SO_4.

References

[1] Kunin R., Meitzner E. G., Bortnick N. // *J. Amer. Chem. Soc.* 84 (1962) 305.

[2] Kunin R., Meitzner E. G., Oline J. A., Fisher S. A., Frisch N. // *Ind. Eng. Chem. Prod. Res. Develop.* 1 (1962) 140.

[3] Kun K. A., Kunin R. // *Polymer Letters*, 2 (1964) 839.

[4] Helfferich F. *«Ion Exchange»* (editor Mc. Graw-Hill), New-York, 1963.

[5] Tager A. «*Physiscal Chemistry of Polymers*», Moscow, Mir, 1972, cap.20.

[6] Paleos J. // *J. Colloid and Interface Science*, 31 (1969) 19.

[7] Frank H. S., Evans M. W. // *J. Chem. Phys.* 13 (1945) 507.

[8] Kuazmann W. «*Advances in Protein Chemistry*» (editors C. B. Anfinson, M. I. Anson, K. Bailey, J. T. Edsall), Academic Press, New York, 14 (1959) 1.

[9] Jones J. H. // *J. Assoc. Offic. Agr. Chemists*. 28 (1945) 398.

[10] Giles C. H., Smith D., Huitson A. // *J. Colloid Interface Science*, 47 (1974) 755.

[11] Fuerstenau D. W. // *J. Phys. Chem.* 60 (1956) 981.

[12] Somasundaran P., Healy T. W., Fuerstenau D. W. // *J. Phys. Chem.* 68 (1964) 3562.

[13] Lewis R. E., Robinson C. P. // *J. Colloid Interface Science*, 32 (1970) 539.

[14] Connor P., Ottewil R. H. // *J. Colloid Interface Science*, 37 (1971) 642.

[15] Shinoda K. «*Colloidal Surfactants*», Academic Press, New York, 1963, P. 229.

[16] Gummings T., Garven H. C., Giles C. H., Rahman S. M. K., Snedden J. G., Stewart C. E. // *J. Chem. Soc.* 62 (1959) 535.

[17] Zollinger J. // *Melliand Textilber*. 37 (1956) 1316.

[18] Zollinger J. // *Helv. Chim. Acta*. 41 (1958) 2242.

[19] Giles C. H., MacEvan T. H. // *J. Chem. Soc.* 70 (1959) 1791.

[20] Tamamushi B., Tamaki K. // *Trans. Faraday Soc.* 55 (1959) 1013.

In: Preparation and Properties of Monomers, Polymers... ISBN 978-1-60021-557-5
Eds: A. Ballada and G. E. Zaikov, pp. 99-110

Chapter 11

How the Structure of Sulphuryl Amides Influences the Light Stabilizing Properties

F. F. Niyazi, O. V. Burykina and I. V. Savenkova
Kursk State Technical University, 94 50 year October Kursk, 305040, Russia

Abstract

Research light stabilizing effect of sulphuryl amide derivatives of carbazole and indan on cellulose diacetate is carried out. It is revealed, that the intensity of light stabilizing properties of sulphuryl amides depends on the degree of conjugation and also substitution of mobile hydrogen in phenylene rings for electron donating substance.

Keywords: cellulose diacetate, carbazole, indan, light stabilizing effect, UV – spectrum.

Introduction

As it is known, aromatic substances with heteroatoms in their structure may cause light stabilizing effect on polymers that are being irradiated. Thus, in order to retrace the process of cellulose diacetate modification we chose qua additions carbazole and indan derivatives, as the latter contain sulphuryl amide groups with heteroatoms of nitrogen, sulphur and oxygen in their composition [1].

Experimental

So, it seemed interesting to study the influence of substitutes in sulphuryl amides concerning light stabilizing effect intensity. For this purpose we modified cellulose diacetate (CDA) with the help of sulphuryl amide derivatives of carbazole and indan.

Among sulphuryl amide derivatives of indan the first to be synthesized was 5-indan sulphonyl chloride - a result of indan and chlorosulphonic acid interaction. Then amines of benzene series were acylation by previously produced 5-indan sulphonyl chloride. Some of the appearing sulphuryl amides came through alkylation by dimethyl sylphate in sulphuryl amide group. Derivatives were obtained by acylating of aniline with special carbazole sulphanilchlorides according to the method [2].

These synthesized sulphuryl amide were used for cellulose diacetate (CDA) modification. The obtained samples were formed as films and radiated by mercury vapour lamp in the air during the period of 24 hours at T = 35^{0}C. Change of viscosity of cellulose diacetate solution before and after irradiation was considered as the index of stabilizing effect of additions – modifiers.

Tables 1 and 2 display the experimental results of sulphuryl amide derivatives of carbazole and indan stabilizing effect.

As it is seen from the data, all the sulphuryl amide derivatives of carbazole and indan redound to photostabilization of cellulose diacetate (CDA).

Studying UV spectrums of carbazole and indan derivatives let us make a supposition about their "shield" mechanism, which causes light stabilizing effect, as sulphuryl amide additions absorb radiation in the same wave length interval as cellulose diacetate.

RESULTS AND DISCUSSION

In order to elicit the light stabilizing mechanism of the examined additions, modified CDA- films were dissolved in acetonitrile before and after being radiated. The obtained solutions were spectrographed in UV (Figures 1-3).

Figure 1. UV spectrum (acetonitrile) 1-CA-film, containing 5 % of compound (II) before irradiation; 2 – the same sample after irradiation; 3 – CA-film, containing 2% of compound (II) before, 4 – and after irradiation.

Measuring UV spectrum displayed, that modifier concentration of 2% to the mass of a fiber change spectrum considerably. It is necessary to note, that after irradiation the spectrum looks differently: absorption bands change to reduction. Possibly, these changes are caused by absorption of CDA decomposition products, which, in its turn, decelerates further polymer decomposition.

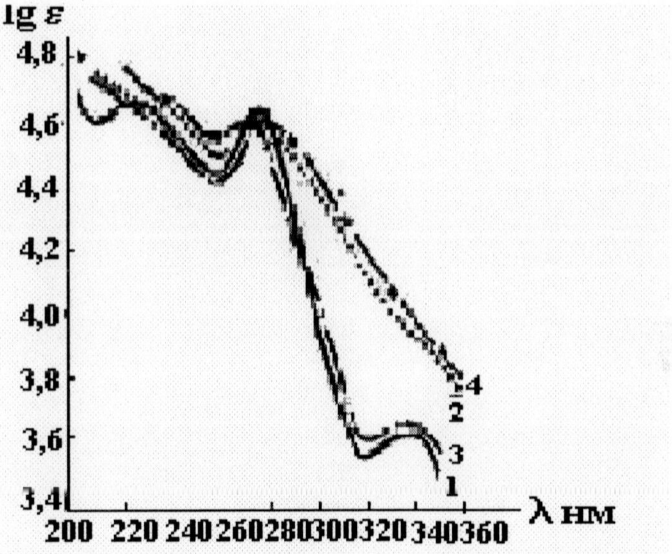

Figure 2. UV spectrum (acetonitrile) 1- CA-film, containing 5 % of compound (III) before irradiation; 2 – the same sample after irradiation; 3 – CA-film, containing 2% of compound (III) before, 4 – and after irradiation.

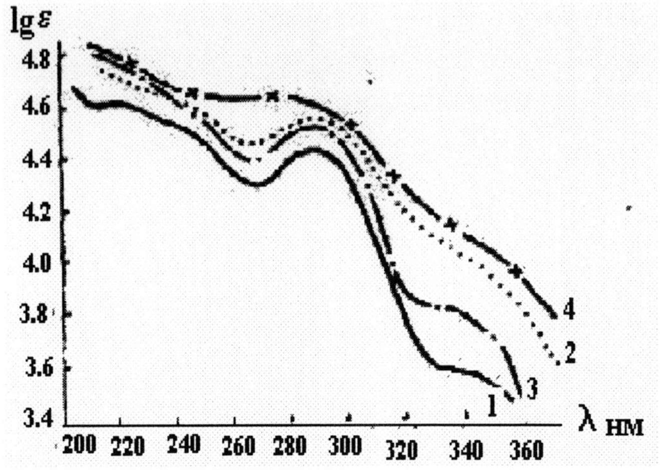

Figure 3. UV spectrum (acetonitrile) 1- CA-film, containing 5 % of compound (V) before irradiation; 2 – the same sample after irradiation; 3 – CA-film, containing 2% of compound (V) before, 4 – and after irradiation.

Table 1. Viscosity maintenance in CDA- films containing indan sulphuryl amide additions in UV radiation

No. of compounds	Formula	Structure	T, ^0C	Viscosity, %
X	$C_{15}H_{15}NO_2S$	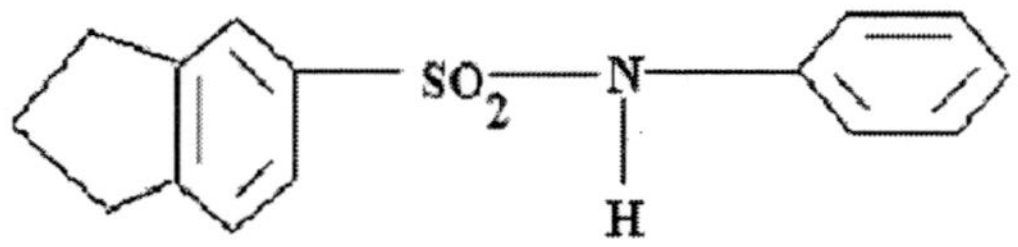	135-136	36,6
XI	$C_{15}H_{15}NO_3S$	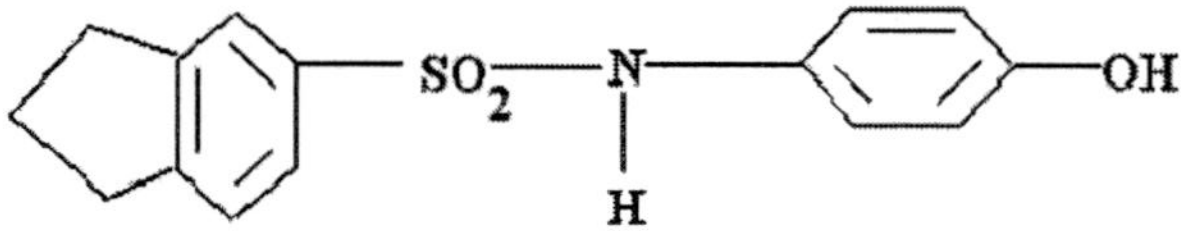	158-159	42,2
XII	$C_{16}H_{17}NO_2S$	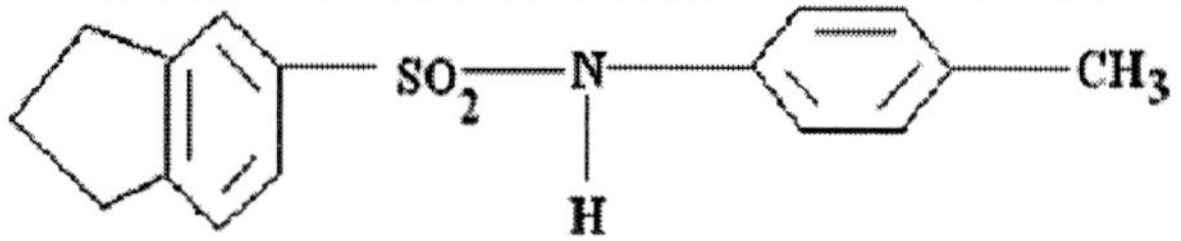	144-146	32,2
XIII	$C_{15}H_{14}N_2O_4S$	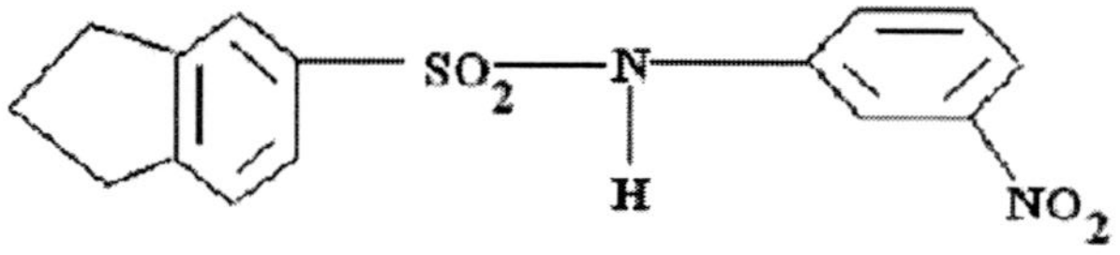	185-186	47,7
XIV	$C_{15}H_{16}N_2O_2S$	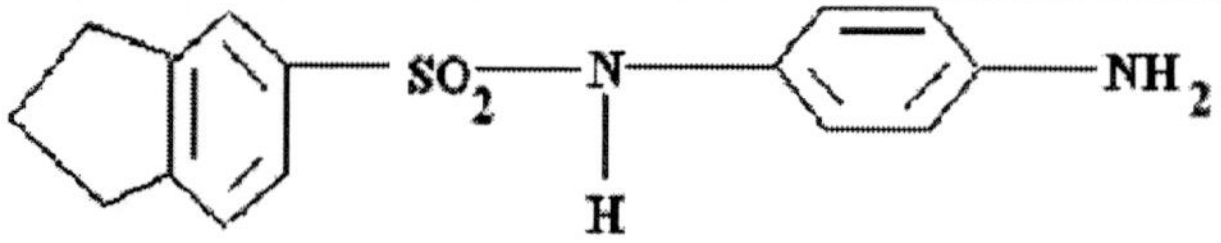	230 destr.	24,0
XV	$C_{24}H_{24}N_2O_4S_2$	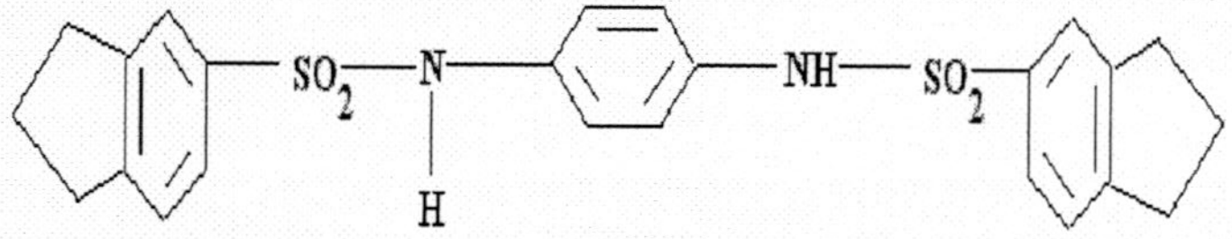	280 destr.	29,1

Table 1. Viscosity maintenance in CDA- films containing indan sulphuryl amide additions in UV radiation (Continued)

No. of compounds	Formula	Structure	T, ^{0}C	Viscosity, %
XVI	$C_{16}H_{17}NO_2S$	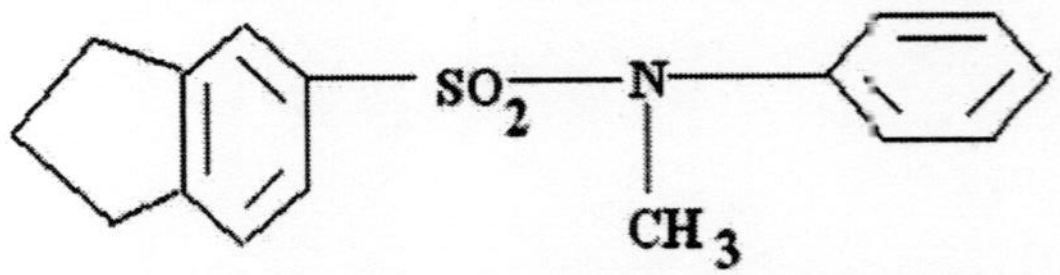	95-96	40,7
XVII	$C_{17}H_{19}NO_2S$	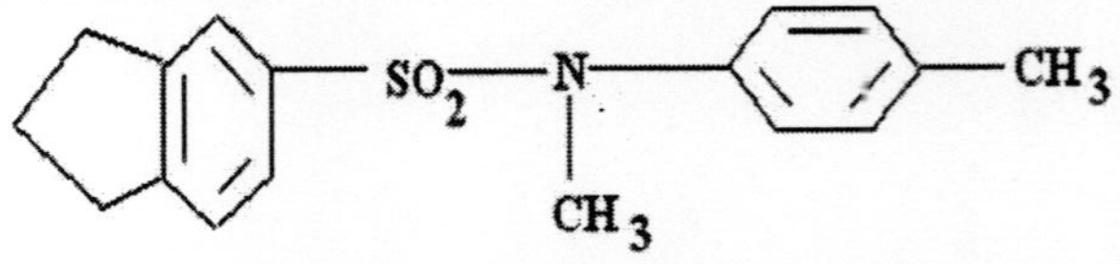	102-103	33,3
XVIII	$C_{16}H_{17}NO_3S$	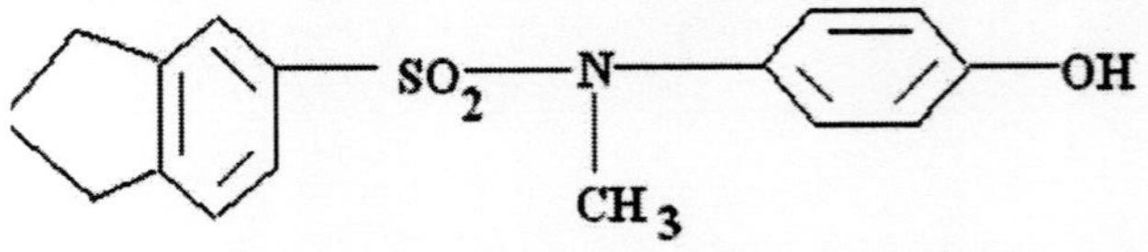	120-121	37,9
XIX	$C_{16}H_{16}N_2O_4S$	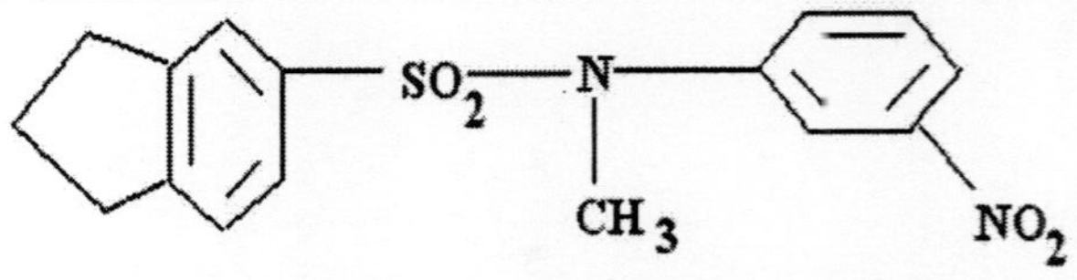	196-197	47,1
XX	$C_{16}H_{18}N_2O_2S$	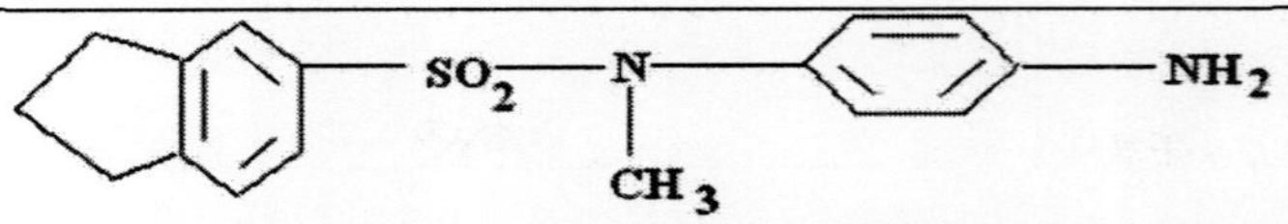	196-197	62,2

Table 2. Viscosity maintenance in CDA- films containing carbazole sulphuryl amide additions in UV radiation

compound No.	Formula	Structure	Viscosity in %		
			Concentration of a stabilizator in % to polymer mass		
			0,5	2,0	5,0
XXI	$C_{13}H_{12}SO_2N_2$	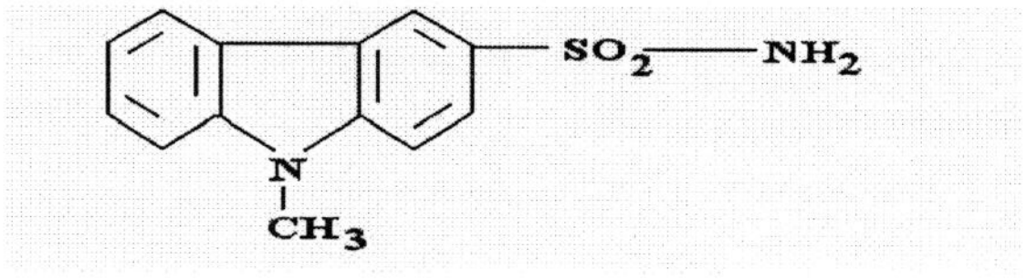	-	63,0	-
XXII	$C_{14}H_{12}SO_3N_2$	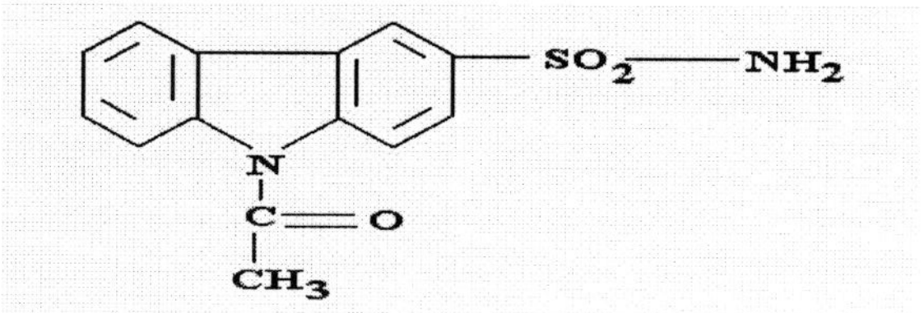	-	74,0	-
XXIII	$C_{20}H_{18}SO_2N_2$	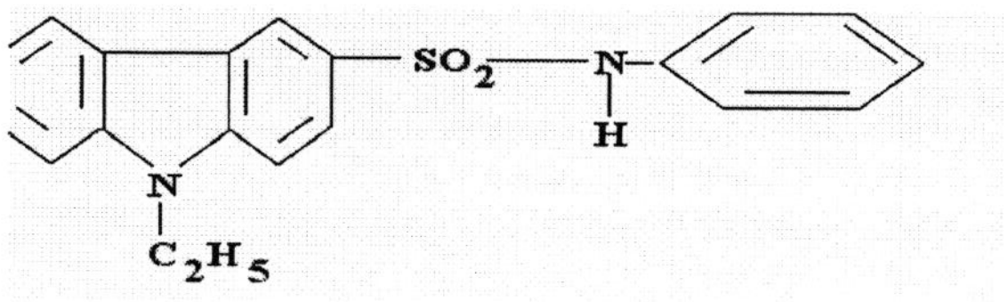	51,0	55,3	64,5

Table 2. Continued

compound No.	Formula	Structure	Viscosity in %		
			Concentration of a stabilizator in % to polymer mass		
			0,5	2,0	5,0
XXIV	$C_{18}H_{13}SO_2N_2Cl$	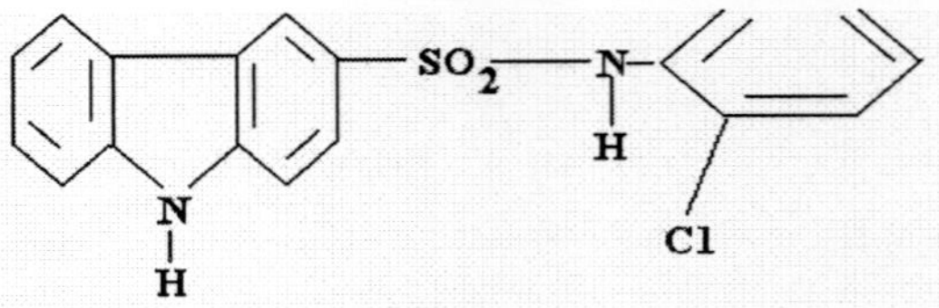	53,5	-	82,9
XXV	$C_{18}H_{13}SO_4N_3$	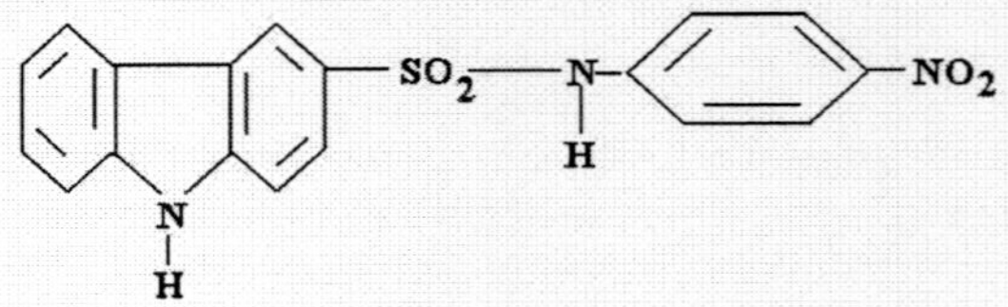	56,9	68,8	83,1
XXVI	$C_{18}H_{14}SO_2N_2$	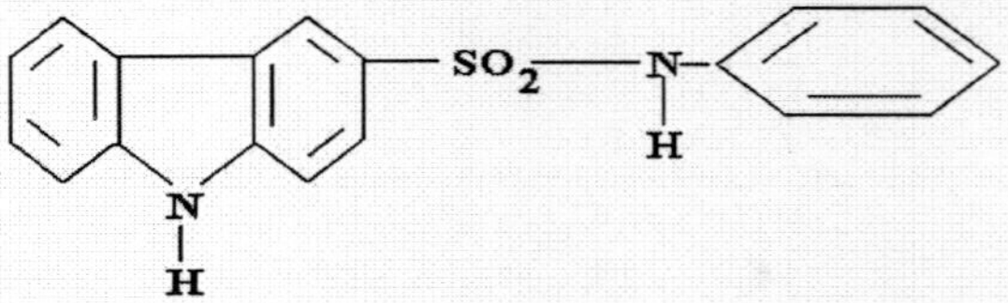	20,5	60,0	72,4
XXVII	$C_{21}H_{20}SO_2N_2$	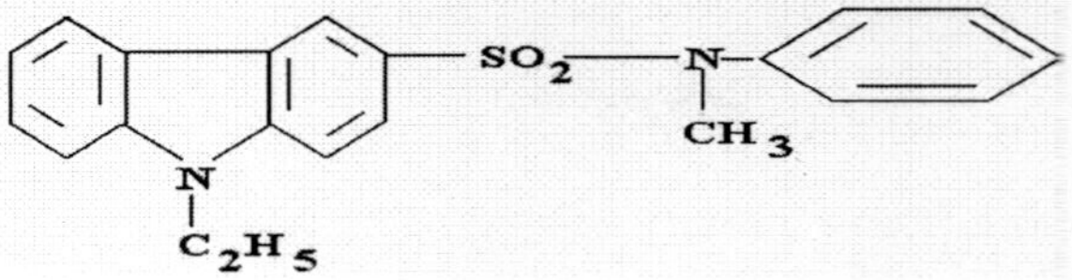	58,4	55,8	62,1

Table 2. (Continued)

compound No.	Formula	Structure	Viscosity in %		
			Concentration of a stabilizator in % to polymer mass		
			0,5	2,0	5,0
XXVIII	$C_{16}H_{18}SO_2N_2$	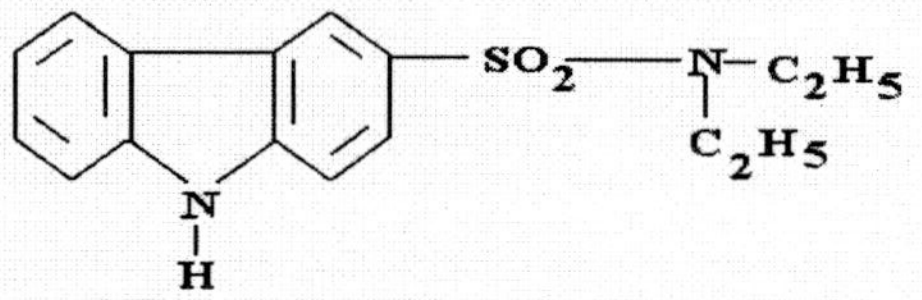	21,9	20,9	47,7
XIX	$C_{24}H_{18}SO_2N_2$	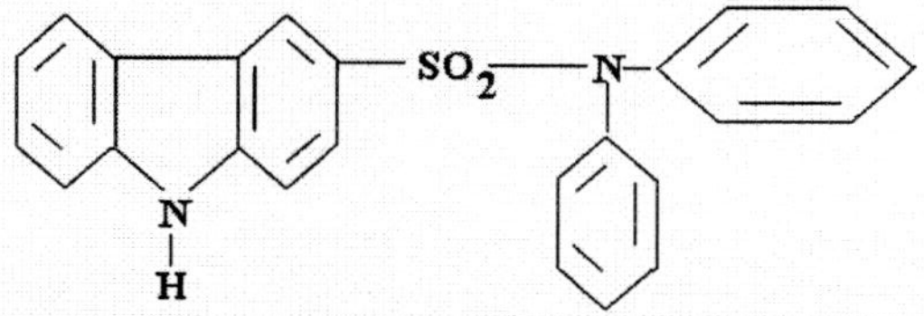	-	42,8	-
XXX	$C_{19}H_{16}SO_2N_2$	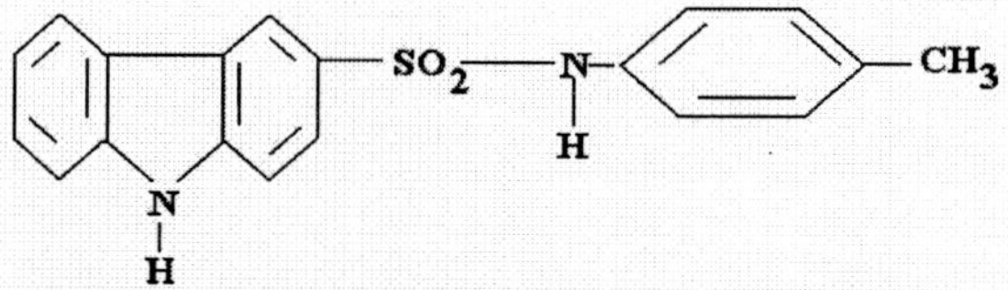	46,2	55,4	59,4
Тинувин-П	-	-	-	34,4	-

As it is established in [3], aromatic sulphuryl amides can exist in solid state in the form of open or cyclic dimers owing to intermolecular hydrogen bonds.

There is an opinion, that sulphuryl amide grouping can have anti-configuration, that is, either O-S-N-H atoms are lying in the same plane or oxygen and nitrogen atoms oppose each other [4]. Thus, as carbazole and indan derivatives include sulphuryl amide grouping, it is possible that light stabilizing effect of additions is due to hydrogen bonds, as it happens in o-hydroxybenzophenone [5].

The light stabilizing effect of the latter is determined by rapid and convertible phototransition of a proton participating in intermolecular hydrogen bond formation.

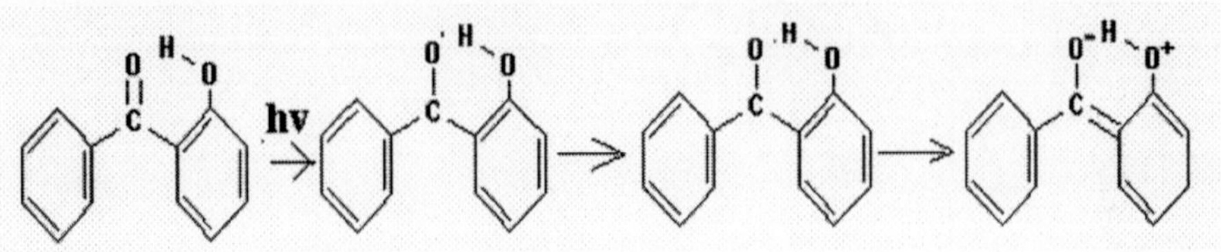

At the same time, there is evidence of arylsulphuryl amides can inhibit radical and chain destruction processes. In order to estimate inhibiting activity of carbazole sulphuryl amide we had to study kinetics of methyl methacrylate free radical polymerization inhibiting, which was initiated by dinitrile azo-diisobutyric acid.

Inhibiting capacity of carbazole sulphuryl amide could be characterized by methyl methacrylate polymerization rate in presence of suggested compounds and in condition of steady flow of the process (W^{st}_{int}), by “gel effect” (W^{gel}_{int}), and also by the fact of inhibiting (F), the value of which is proportionate to inhibiting rate constant.

It is quite conspicuous in kinetic parameters in Table 3, all carbazole sulphuryl amides reduce to some extent polymerization rate.

Table 3. Kinetic parameters methyl methacrylate polymerization with carbazole sulphuryl amides

Compound No.	$W^{st}_{int}10^4$, mole/l s	F_{st}	$W^{gel}_{int}10^4$, mole/l s	F_{gel}
Without Inhibitor	6,9	-	62,0	-
XXIII	5,0	0,64	41,3	0,83
XXV	6,0	0,28	42,2	0,79
XXVII	3,8	1,28	23,0	2,32
XXVIII	3,5	1,49	32,8	1,36
XXX	4,7	0,77	34,7	1,32

However, absence of the induction period implies that these compounds are weak inhibitors. Possibly the accessorial protective effect of carbazole sulphuryl amide additions is due to their reducing role in radical process of polymer photodestruction.

Adding in various substituents into sulphuryl amide group has significant influence on their stabilizing activity (as seen in Tables 1 and 2). In case of XIV and XX indan sulphuryl amide derivatives we can see a sharp difference in light protection effect (24,6 and 62,6 % correspondingly). Though these compounds differ only by methyl group connected with nitrogen from sulphuryl amide grouping. This difference is proved by UV spectrum in stabilized CDA-films, irradiated during 24 hours and non irradiated (Figures 4-5).

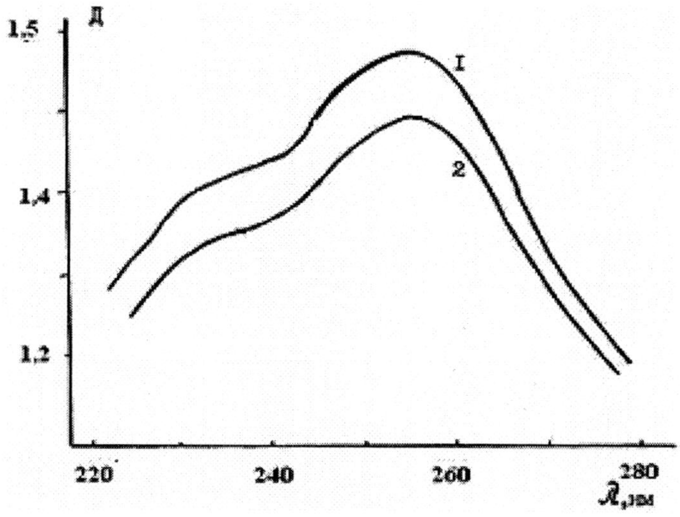

Figure 4. UV spectrum of absorption in CDA - films, containing 2% of XX indan sulphuryl amide to polymer mass: 1- CDA - film before irradiation; 2-after 24 hours UV – irradiation.

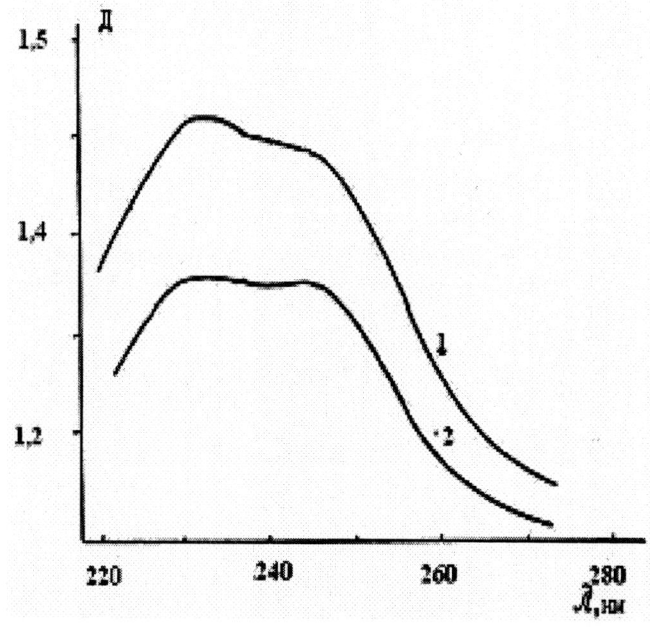

Figure 5. UV spectrum of absorption in CDA - films, containing 2% of XIV indan sulphuryl amide to polymer mass: 1- CDA - film before irradiation; 2-after 24 hours UV – irradiation.

Thus, cellulose diacetate CDA, modified by XX indan sulphuryl amide has its maximum of absorption at 256 nm, while CDA containing XIV compound has it at 232 nm, which deposes stronger conjugation effect in case of XX. Besides, in irradiated CDA -films ratio of absorption intensity to some initial value at corresponding λ_{max} in XIV compound is 0,68, while in XX it is 0,86. That implies, that being irradiated XIV stabilizer is expended 1,3 times as quick as XX.. Apparently, methylized nitrogen in sulphuryl amide group stabilize the structure, which, in its turn, increases the intensity and length of light stabilizing effect.

Comparison of experimental data concerning light stabilizing activity in indan sulphuryl amides with the results occurring in carbazole derivatives displays that changing indan fragment to carbazole cycle we increase conjugating effect and, as a result, increase in light stabilizing activity for 15-26% (Table 4).

Table 4. The comparative characteristics of stabilizing effect of sulphuryl amide derivatives of carbazole and indan

Compound	Structural formula	Viscosity, %
indan derivatives (X)	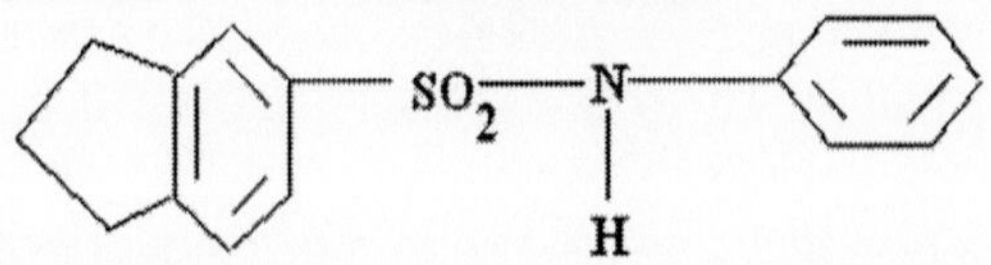	36,6
carbazole derivatives (XXVI)	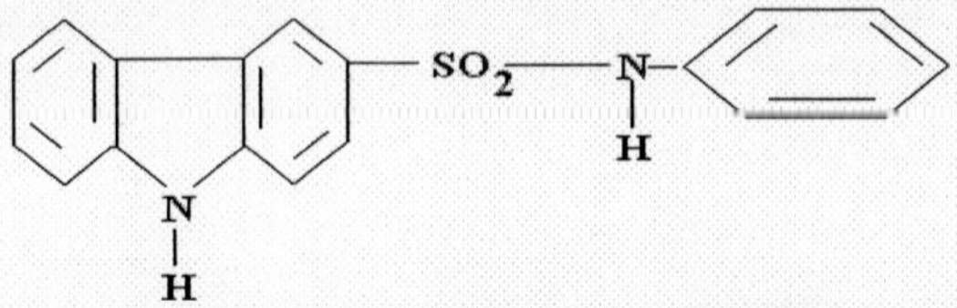	72,4
indan derivatives (XII)	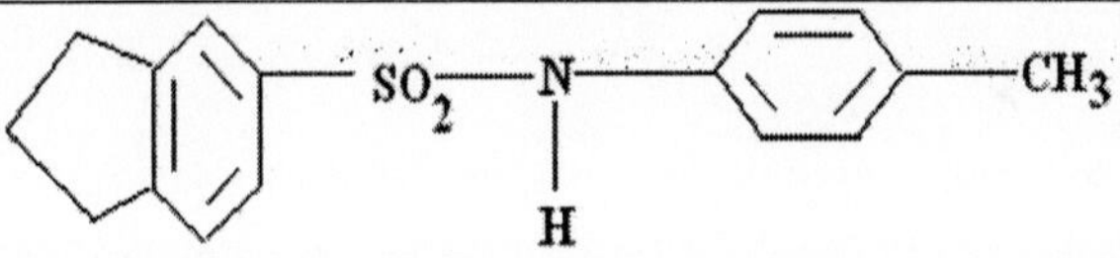	32,2
carbazole derivatives (XXX)	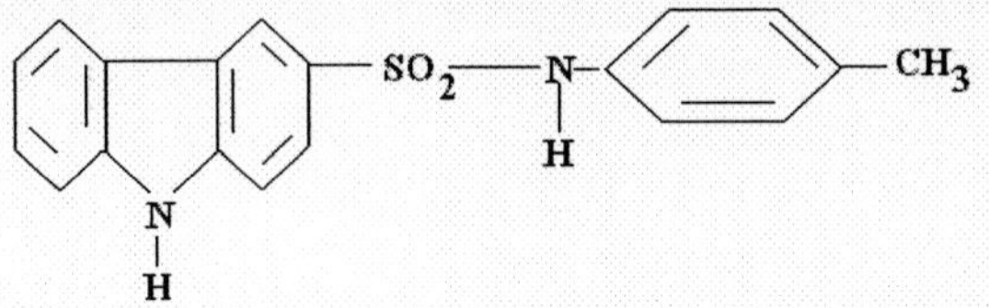	83,1
indan derivatives (XVI)	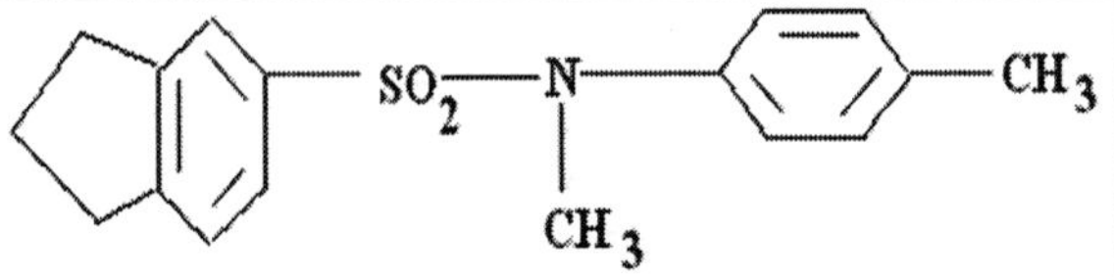	40,7
carbazole derivatives (XXVII)	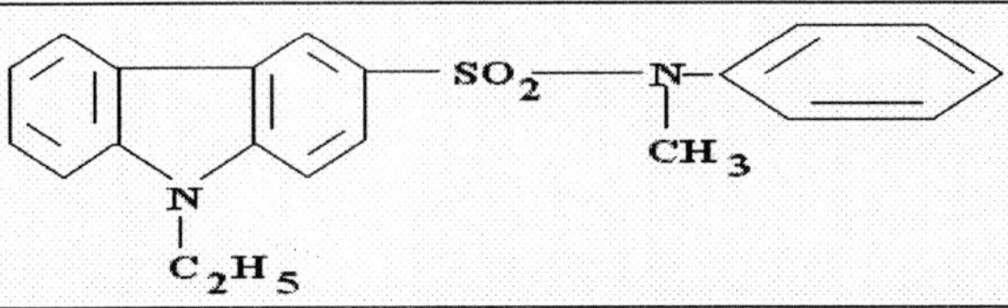	62,1

Substitution of hydrogen for alkyl residue in heteroatom of carbazole cycle causes high level photoprotective effect even at 0,5% stabilizer concentration in polymer, the more so, this effect doesn't change considerably with further increasing of concentration (Table 3 XXVI and XXIII compounds).

Inhibiting influence undergoes notable change depending on substitute in both carbazole cycle, and in amine component. Thus, carbazole sulphuryl amides derived from aliphatic amines (XXVIII, XXIII, XXII) show lower activity, than those of acrylamin derivatives (XXVIII, XXVII).

Substitute intrusion into phenylen rings results in sharp increase of light stabilizing activity in carbazole derivatives, as well as in those of indan (XXVI and XXX compounds or XVI and XVIII compounds). Thus, intruding electron accepting substitutes (-NO_2) causes greater stabilization of CDA, than penetration of electron donation particles (-CH_3, -OH, -NH_2). Possibly, it is connected with ability of electron accepting particles to draw forward electronic density of benzene ring, as a result, carbon atoms get positive charge and hydrogen connected with them becomes mobile.

Comparing of stabilizing activity among electron donation substitutes allows us to notice, that the stronger electron donating properties are exposed, the greater is light stabilizing effect (compound XX – substitute –NH_2 (62,2%); compound XVIII – substitute –OH (37,9%), compound XVII- substitute –CH_3 (33,3%).

It is important to notice, that sulphuryl amides derivatives of both indan and carbazole have a number of valuable properties of practical application: they are resistant to thermal shock (T decomp =120-300^{0}C); non-toxic (LD_{50}=1000 mg/kg); well consistent with CDA in common solvents; acquire small volatility.

Conclusion

The results we have obtained give us grounds to consider, that the intensity of light stabilizing properties of sulphuryl amides depends on the degree of conjugation and also substitution of mobile hydrogen in phenylene rings for electron donating substance.

References

[1] Foigt, Stabilization of Synthetic Polymers against Influence of Warmth and Light. Leningrad: *Chemistry*, (1972), 135 (in Russian).

[2] Abdirazakov, F. Niyazi, O. In, I. Kolontarov, V. Shishkina- All-Union Conference. in *Textile Fabric Refinement*, Thesis of Conference, Dushanbe, (1980), 88, (in Russian).

[3] A.E Lutskiy, I.K. Ishchenko. *Organic Chemistry Journal* (1968), 38 vol.8, 1629-1635, (in Russian).

[4] A.E Lutskiy, I.K. Ishchenko.-*Organic Chemistry Journal* (1968), 38 vol.8, 1618-1629, (in Russian).

[5] Burgess A.E., *Nat. Bur. Stand, Circ.* 525, 149, (1953).

In: Preparation and Properties of Monomers, Polymers… ISBN 978-1-60021-557-5
Eds: A. Ballada and G. E. Zaikov, pp. 111-118

Chapter 12

Of Complex Aerohydrodynamic Research and the Effectiveness of Arresting Dispersed Particles for Barbotage – Rotation

A. K. Panov, R. R. Usmanova, V. G. Zaikov [*] ***and G. E. Zaikov*** [*]
Branch UGNTU, 2 Oktyabrya bulvard, Sterlitamak 453118, Bashkortostan, Russia
[*]N.M. Emanuel Institute of Biochemical Physics, RAS,
4, Kosygin Street, Moscow 119991, Russia

Abstract

The design of barbotage-rotation installation with different profiles of blades of the rotator has been created on the basis of modeling method.

As a result of the complex of aerohydrodynamic research in the system "gas - liquid" carried out on the above unit certain recommendations on the improvement of the construction and operating conditions have been developed with a view to a rise in the effectiveness of dust removal.

Keywords: aerohydrodynamic, dispersed particles, barbtage – rotation , profiles of blades, modeling method, improvement of the construction, operating conditions.

Conditional Designations

E – loss efficiency of curling
G – gas expenditure, m^3 / s
K – curling efficiency, k = (0,2 ÷ 1,0);
n – experiment constant;
Q – liquid expenditure, m^3 / s;
R – apparatos radius, m;
r-rotation radius ,m
V_1 – gas speed on inlet, m / s;

V_2 – gas speed on inlet, m / s;
P_g – gas density, kg / m^3;
P_l – liquid density, kg / m^3;

Barbotage-rotation installations designed for clearing and cooling gas discharges are applied in chemical, oil refining, and other fieds of industry.

The above units are important in solving ecological problems, e.g. the refining process of waist (exhaust) gas removed from the unit into the atmosphere. [1]

That's why research on the improvement of the design and operating conditions of barbotage-rotation units has always been urgent and remains problem of today. [2]

With a view to intensification of the refining process and a rise in the effectiveness of the barbotage-rotation unit, a complex of aerohydrodynamic research on the improvement of the construction of the rotator and operating conditions of the unit has been carried out.

1. Description of the Unit for Dust Removal Process Research and the Technique (Method) of the Experiment

Research on aerohydrodynamics of curled (rotated) flows and the effectiveness of dust removal in the system "gas - liquid" was carried out on the barbotage-rotation unit with different profiles of blades of the rotators. The unit was designed taking into consideration possible approximations to real conditions of industrial installations. (Figure 1)

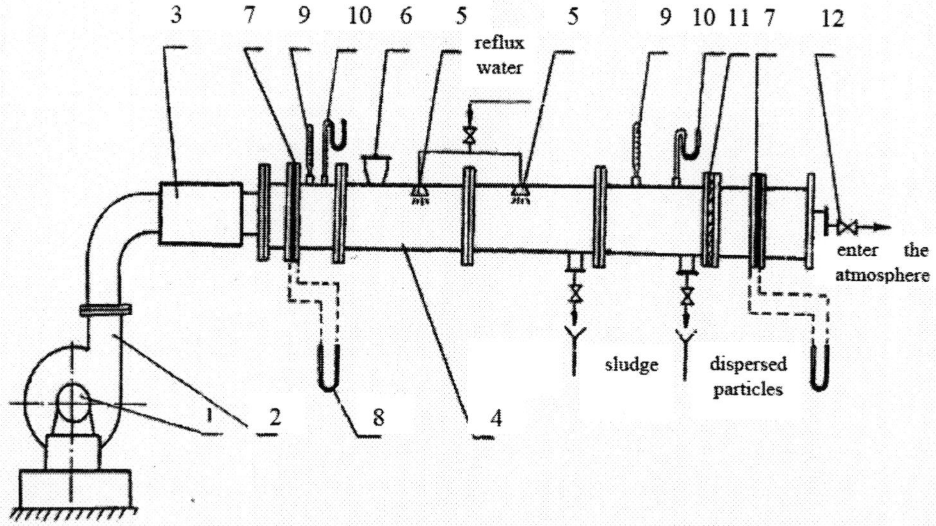

1 – electric motor; 2 – ventilator (fan); 3 – electric heater; 4 – rotation chamber (cell); 5 – sprayers; 6 – bunker; 7 – diaphragm; 8 – pressure – gauge; 9 – thermometer; 10 – manometer; 11 – filter; 12 – valve.

Figure 1. Flow diagram of the experimental unit.

It was taken into account that the rotator should ensure not only a high degree of dust removal but be constructively well coordinated with the barbotage – rotation device. Air and white soot (smoke - white) with the particle size of d = 2 ÷ 40 mkm were used as a model system.

Atmospheric air forced by ventilator 2 (VC 1446, producer being Industrial Ventilation, Sterlitamak, Bashcortostan) was passed throngh electric heater 3 at the required heat temperature (100^0C). Dusty air rate of flow $G=(0,6 \div 4,0) \cdot 10^{-2}$ m^3/s. After reaching the operating temperature conditions and sprayers 5, placed in front of and behind the rotator, the flow was given reflux liquid, reflux feed being regulated by rotameter RS – 7. Simultaneously the flow was given white soot from feed bunker 6. While the dusty air was passing through the blade rotator, in the field of centrifugal forces dust removal and cooling of the gas flow took place. The unremoved dispersed particles of the flow fell out in filter 11. The removal of the sludge and dispersed particles was carried out through unions in the bottom of the device.

The purified air was discharged into the atmosphere.

Sprayers 5 were placed in front of and behind the atmosphere rotator for the improvement of purifying conditions.

Reflux feeding was regulated by rotameter RS – 7 ($Q=(5,5 \div 14) \cdot 10^{-4} m^3/s$).

The liquid reflux favoured the enlargement of the unremoved, dispersed particles and the reduction of time of removal. The sludge formed was washed off with the liquid sprayed by the sprayer placed behind the rotator and dischared into the sludge – collector.

The subject of the Research was constructions of rotators with four kinds of profiles of blades: parabolic, hyperbolic, sine wave and direct (Figure 2).

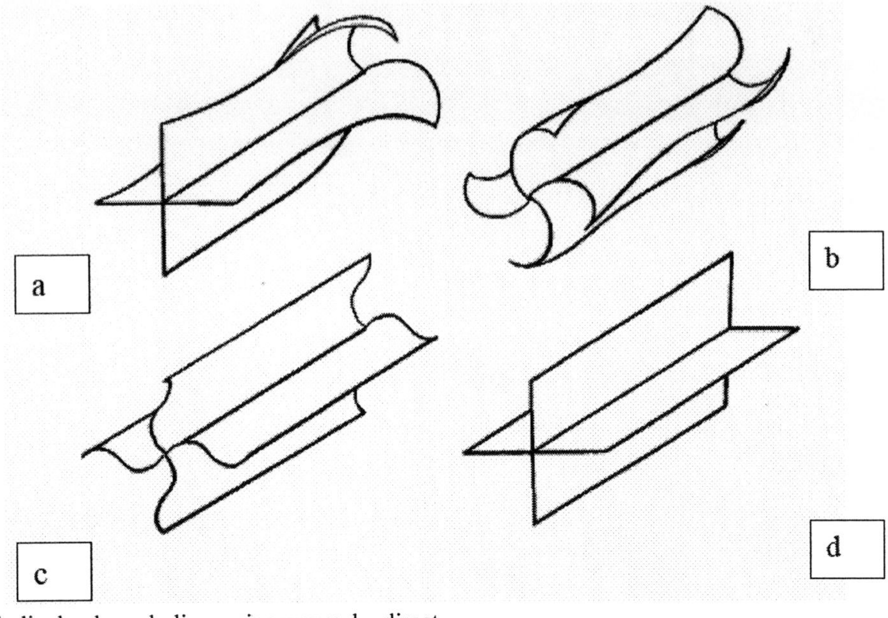

a – parabolic; b – hyperbolic; c- sine wave; d – direct.

Figure 2. Profiles of blades of rotators.

2. Evaluation of the Effectiveness of Dust Removal

Full (complete) effectiveness of dust removal was (calculated) expressed in percentage [3] as a ratio between mass of removed dust and mass of initial dust.

$$\eta = \frac{C_n - C_k}{C_n} = \frac{C_{rem}}{C_n} \cdot 100\%, \tag{1}$$

where C_n – mass expenditure of rigid particles on exit, kg / s; C_{rem} – mass expenditure of removed particles, kg / s.

According to the results of the effectiveness accouts there were built diagrams for different profiles of blades of the rotator. As the data obtained shows for the models of blades of the rotator being under research the main mass of particles is removed completely. The effectiveness of dust removal remains within 85 – 99 %. With high reflux common efficiency of the device rises (Figure 3). With minimum reflux of Q = 8,5 · 10^{-4} m^3/s, common efficiency comes to 85 %.

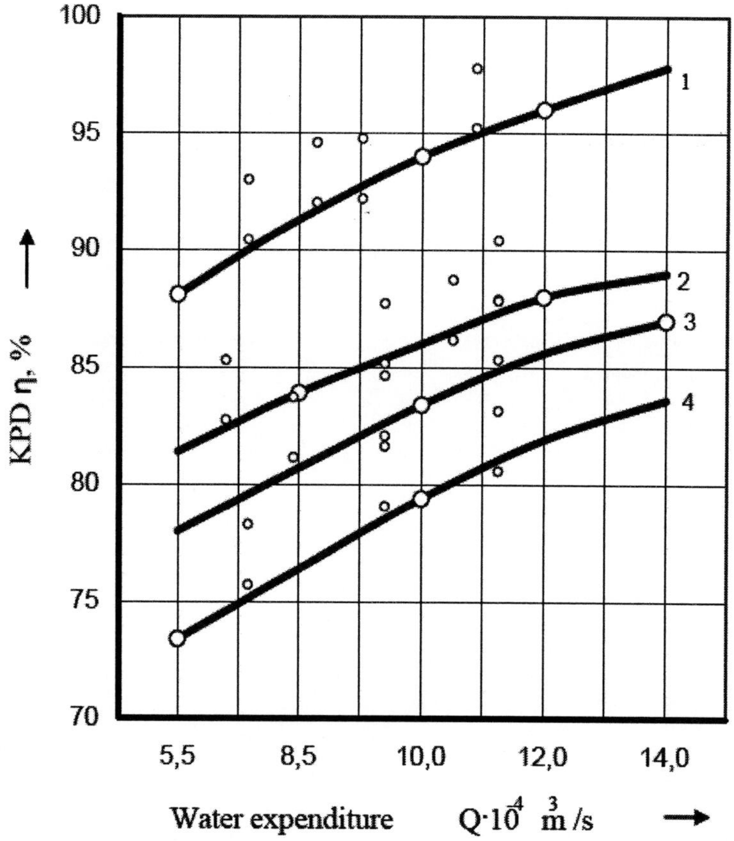

1 – parabolic; 2 – hyperbolic; 3 – sine wave; 4 – direct.

Figure 3. Dependence of the device efficiency on water expenditure at constant air expenditure (G = 40 · 10^{-3} m^3 / s) for different profiles of the rotator blades.

The data obtained shows that even with little reflux a high degree of dust removal is obtained. Research shows that in the investigated range the initial dust concentration practically does not influence the effectiveness of dust removal. This is of great importance for the device work in industrial service conditions where the dust content being the function of the technological process conditions may vary over a wide range.

With high air expenditure the device efficiency falls (drops) (Figure 4). This can be explained by the fact that at significant air expenditure and little water expenditure some rigid particles have no time to enter the contact with the liquid and are discharged from the device.

In the diagram points (dots) show experimental efficiency values, continuous lines show theoretical values calculated by formula (1).

It is interesting to note that the least dust removal efficiency is obtained for rectangular profile of blades, coming to 85 %.

The highest device efficiency (up to 99 %) is obtained for parabolic profile of rotator blades.

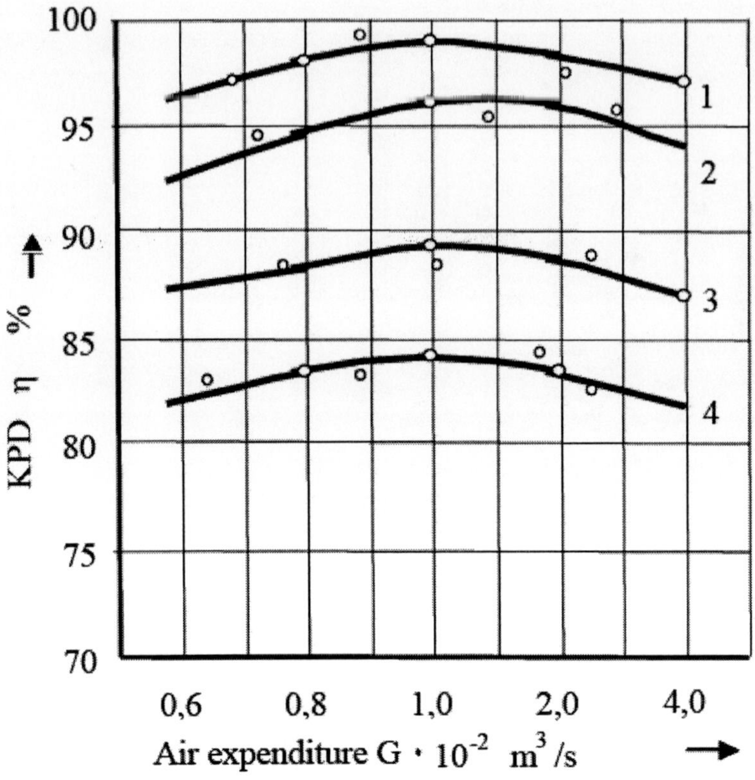

1-parabolic; 2- hyperbolic; 3- sine wave; 4 – direct.

Figure 4. Dependence of the device efficiency on air expenditure at constant water expenditure (Q=8,5 · $10^{-3}m^3/s$) for different profiles of the rotatot blades.

3. Aerohydrodynamic Research Results for Barbotage – Rotation Device in Comparison with Theoretical Ones

There is a general tendency to a rise in hydrolic resistance with high speed in a free cut.

The maximum value of hydrolic resistance at specific reflux of $Q = 14 \cdot 10^{-4}$ m^3 / s comes to $\Delta P = 30$ Па for hydrolic profile of blades, $\Delta P = 20$ Па for direct and sine wave ones.

Thus constructive modifications of the profiles of the rotator blades enable one to raise significantly dust removal effectiveness (Figures 3, 4.) with slight (small) hydrolic resistance increase.

Curling efficiency of rotator K varies from 0,2 ÷ 1,0. The experimental data analysis shows that hydrolic resistance efficiency ξ at some value K does not depend on liquid expenditure which is explained by two factors connected with reflux feeding of the device.

On the one hand reduction of ξ is observed because of gas tangential speed drop on the account of brake action of the liquid, on the other hand increase of ξ is connected with a rise in gas flow pressure on the liquid carrier.

On this foundation ξ account technique was developed including ξ account formula for the "dry" device

$$\xi_{dry} = \frac{1}{n}((R/r)^{2n} - 1) + \frac{1}{K^2}\left(\frac{V_2}{V_1}\right)^2, \tag{2}$$

the formula of loss account of gas pressure on the liquid carrier

$$\xi_{tr} = 4 \cdot \left(\frac{Q}{G}\right)^{0,6} \sqrt{1 + \frac{1}{K^2}}, \tag{3}$$

and the final formula for ξ account of the reflux fed rotation device

$$\xi_{sp} = \frac{1}{n}((R/r\ )^{2n} - 1) + \frac{E^2}{K^2}\left(1 + \frac{\rho_l}{\rho_g}\right) \cdot \left(\frac{V_2}{V_1}\right)^2 + 4\left(\frac{Q}{G}\right)^{0,6} \sqrt{1 + \frac{1}{K^2}}, \tag{4}$$

Taking into consideration the presence of a disperse phase and partial loss of the flow curling. The ξ accounts by this formula will be well coordinated with the experiment data (Figure 5).

In the diagram theoretical values of ξ are marked by a continuous line, the experiment data is marked by points (dots). The relative divergence of theoretical values calculated by formula (4) and the experimental data made no more than 15 %.

Thus the technique offered with an adequate accuracy of accounts can be recommended for aerohydrodynamic research of the flows in the barbotage – rotation device with different constructive parameters of the rotator.

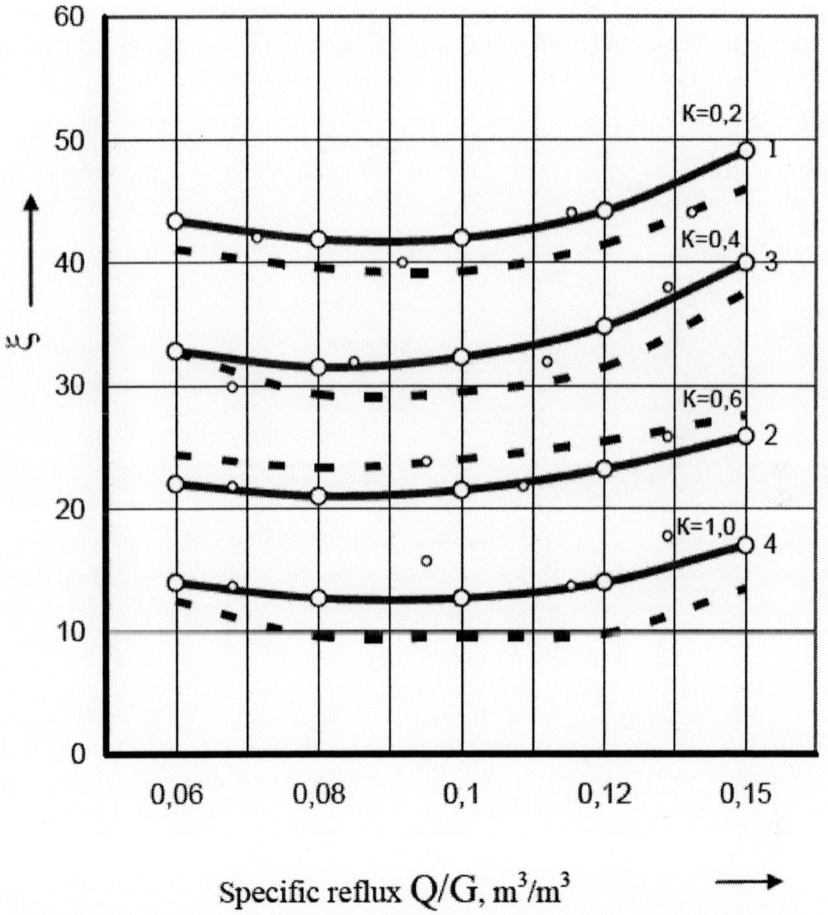

Figure 5. ξ dependence on specific reflux of the device with different values of curling efficiency.

Conclusion

1) Taking into consideration research results on dust removal from gas it is advisable (expedient) in what follows to use the barbotage – rotation device with parabolic profiles of rotator blades, having high effectiveness of dust removal (up to 99 %) and small hydrolic resistence, as an optimum variant in dust removal system.
2) The introduction of barbotage – rotation device with parabolic profile of rotator blades is offered to Bashcortostan refinery plants for dust removal and cooling gas discharges.

References

[1] Patent 2182843 RF MKI V 01 D 47 / 06 *Barbotage – rotation device* / A.K. Panov, R.R. Usmanova, V.M. Titov et al. – N. 200012 / 12; requested 27.11.00; published 27.05.02.

[2] Panov A.K., Usmanova R.R., Minsker K.S. Effectiveness raise of the process of clearing furnace combustion gases / *Chemical Industry* / - 2003 N. 4. P. 19 – 24.
[3] Ugov V.N., Valdberg A.J. Preparation of industrial gases for clearing. – M.: *Chemistry*, 1985. – P. 206.

In: Preparation and Properties of Monomers, Polymers... ISBN 978-1-60021-557-5
Eds: A. Ballada and G. E. Zaikov, pp. 119-131

Chapter 13

THE MECHANISM OF SELECTIVE OXIDATION OF ETHYLBENZENE WITH DIOXYGEN INTO α-PHENYLETHYLHYDROPEROXIDE AT CATALYSIS BY FE(III)(ACAC)$_3$, ACTIVATED WITH ADDITIVES OF 18-CROWN-6 AS LIGAND-MODIFIER

L. I. Matienko* and L. A. Mosolova
N. M. Emanuel Institute of Biochemical Physics, Russian Academy of Sciences, 4, Kosygin str., Moscow, 119991 Russia

ABSTRACT

The kinetic investigation of the oxidation of ethylbenzene with dioxygen catalyzed by Fe(III)(acac)$_3$ in the presence of additives of macrocyclic polyether 18-crown-6 (18C6) was performed. The principle possibility of the composition catalysts construction including complexes Fe(III)(acac)$_3$ and 18-crown-6 (18C6) as ligand-modifier for the control the selective ethylbenzene oxidation with dioxygen into α-phenylethylhydroperoxide (PEH) is demonstrated. The value method of the activity of the iron modifier catalyst (Fe(II)(acac)$_2$)$_p$ (18C6)$_q$ in micro steps of the chain initiation (activation O$_2$) and chain propagation (Cat+RO$_2^{\bullet}$→) at the initial stages of the chain-radical ethylbenzene oxidation is applied. The proposed "dioxygenase-like" mechanism of the transformation of the (Fe(II)(acac)$_2$)$_p$ (18C6)$_q$ complexes into more active catalytic particles in the course of the oxidation of ethylbenzene is discussed.

Keywords: oxidation, ethylbenzene, α-phenylethylhydroperoxide, homogeneous catalysis, dioxygen, iron (II or III) (acetylacetonates), 18-crown-6.

---

* E-mail:matienko@sky.chph.ras.ru

1. INTRODUCTION

The selective oxidation of hydrocarbons into hydroperoxides, primary products of oxidation is the most difficult problem because of the high catalytic activity of the majority of applied catalysts in ROOH decomposition. At the same time, the problem of selective oxidation of alkylarens (ethylbenzene and cumene) with dioxygen in ROOH, is of current importance from practical point of view in connection with ROOH use in large-tonnage productions such as production of propylene oxide and styrene (α-phenylethylhydroperoxide), or phenol and acetone (cumyl hydroperoxide) [1].

At recent decades the interest to fermentative catalysis and investigation of possibility of modeling of biological systems able to carry out selective introduction of oxygen atoms by C–H bond of organic molecules (mono- and dioxygenase) is grown. Unfortunately, dioxygenase able to realize chemical reactions of alkane's dioxygenation are unknown [2].

The method of homogeneous catalysts modifying by additives of different electron-donor ligands for increasing of selectivity of liquid-phase oxidation processes was proposed by us. We established the mechanism of catalysis by nickel complexes $Ni(II)(L^1)_2$ (L^1=acac$^-$ (acetylacetonate ion)), activated by additives of electron-donor monodentate ligands-modifiers L^2 (L^2=HMPA, DMFA, NMP (N-methylpirrolidone-2), MSt (M=Li, Na, K)) in the process of selective oxidation of alkylarens (ethylbenzene and cumene) with dioxygen into corresponding hydroperoxides and proposed the effective methods for control of the selective oxidation of ethylbenzene into α-phenylethylhydroperoxide (PEH) at catalysis by $Ni(II)(L^1)_2$ [3-7]. Some of the most effective methods are connected with the use of the additives of quaternary ammonium salts R_4NBr and crown-ethers (L^1=acac$^-$, enamac$^-$ (enaminoacetonate ion)) as L^2 exo ligands-modifiers. In these cases the control of the catalytic activity of $Ni(II)(L^1)_2$ by R_4NBr- or crown-ethers (L^2) additives consists in the increase of activity of primary complexes $Ni(II)(L^1)_2\cdot(L^2)_n$ as selective catalysts and stabilization of new selective catalysts, complexes of the common formula $Ni(II)_x(L^1)_y(L^1_{ox})_z\cdot L^2$ (L^1_{ox}= $(MeCOO)^-$ or $(NHCOMe)^-$) “A”, the intermediate products of transformation of primary complexes in the course of the oxidation of ethylbenzene in consequence of oxidative cleavage of L^1 ligand under the action of O_2 [8-10]. Recently we established that $Fe(III)(acac)_3$ was not enough effective catalyst for the selective ethylbenzene oxidation into PEH as compared with Ni(II) or Co(II) acetylacetonate complexes [11]. The selectivity of oxidation of ethylbenzene into PEH in the presence of $Fe(III)(acac)_3$ varies only slightly over a wide range of Cat concentrations ([Cat](0,5 - 7,5)$\cdot 10^{-3}$) mol/l), being equal to 45-65%. In the presence of $Fe(III)(acac)_3$ unlike complexes of Ni(II) and Co(II) the products of the oxidation of ethylbenzene acetophenone (AC) and methylphenylcarbinol (MPC) like PEH are the major products and are formed parallel to PEH formation with high rates ($w_P/w_{PEH} \neq 0$ at $t\rightarrow 0$, P=AC, MPC), AP and MPC are formed also parallel ($w_{AP}/w_{MPC}\neq 0$ at $t\rightarrow 0$) mainly at the micro step of chain propagation Cat + $RO_2^{\bullet}\rightarrow$ as at the beginning of reaction, so at deeper stages of oxidation [11]. We showed that it was possible to increase the selectivity of the oxidation of ethylbenzene into PEH at the catalysis of $Fe(III)(acac)_3$ using the additives of exoligands-modifiers L^2 (HMPA or quaternary ammonium salts R_4NBr) [12,13]. It was established that the increase in selectivity at the beginning of the oxidation at the catalysis by the $Fe(II)(acac)_2$ L^2 complexes [L^2 = HMPA] formed in the course of ethylbenzene oxidation was explained by the change of the ratio of the rates of chain initiation (w_i, activation O_2) and chain propagation (w_{pr}, Cat+$RO_2^{\bullet}$→), w_i / w_{pr} [12]. In the case of use of quaternary ammonium salts R_4NBr as exoligands-modifiers L^2 the significant increase in PEH selectivity in the course of ethylbenzene oxidation is connected with transformation of $(Fe(II)(acac)_2)_x$ $(R_4NBr)_y$ under the action of molecular O_2 into new catalytic particles [13]. In this case the selective catalysis by new catalytic particles is connected with the hindering of the heterolytic decomposition of PEH, and of the methylphenylcarbinol and acetophenone formation in the parallel with PEH micro steps (Cat+$RO_2^{\bullet}$→). The probable "dioxygenase-like" mechanism of the transformation of iron catalyst (as well as in the case of use of DMFA as ligand - modifier) into more active catalytic particles in the course of oxidation process was proposed by analogy to catalysis by acetyl acetone dioxygenase (M = Fe(II)) (Dke 1), which promotes dioxygen-dependent conversion of 2,4-pentandione into acetate and methylglyoxal (Scheme 1) [14].

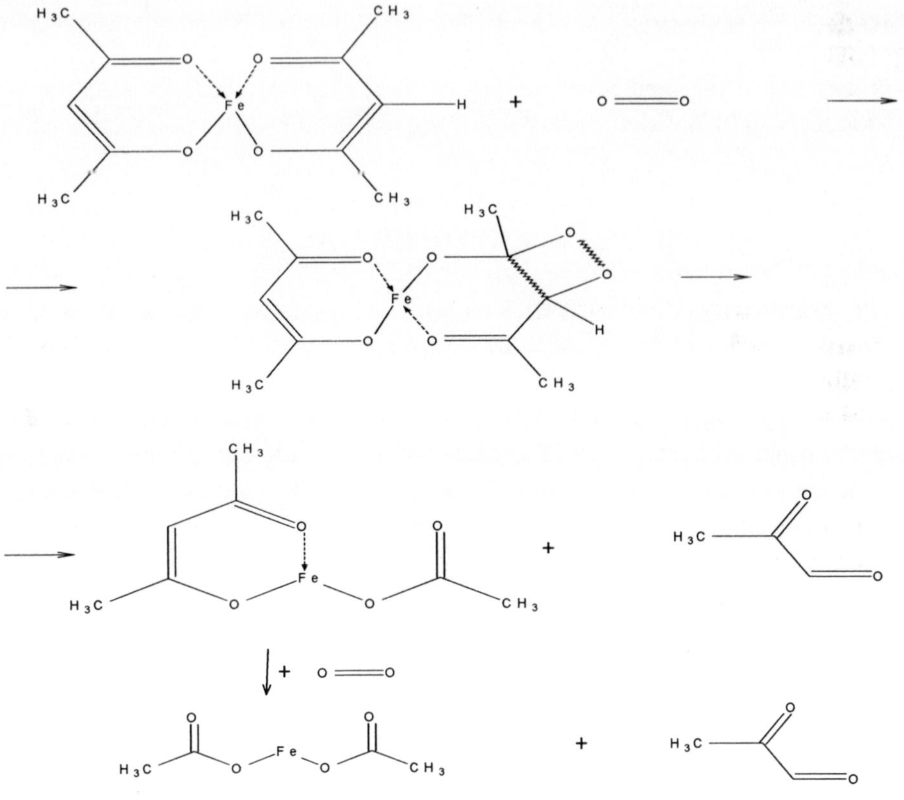

Scheme 1. The principle scheme of dioxygen-dependent conversion of 2,4-pentandione catalyzed by acetyl acetone dioxygenase Fe(II).

The active selective catalyst is the intermediate product of the transformation, hetero ligand complex of the common structure:

$Fe(II)_x(acac)_y(OAc)_z(L^2)_n$ (L^2= R_4NBr, DMFA) [12,13].

With the formation of the final product of the conversion of acetylacetonate ligands, Fe(II) acetate, is connected obviously the decrease in S_{PEH}, the catalysis of heterolysis of PEH into phenol and acetaldehyde.

On the basis of literature data we may expect for the increase in S_{PEH}^{max} and conversion degree C of the oxidation of ethylbenzene catalyzed by $Fe(III)(acac)_3$ in the presence of additives of 18K6 by analogy with catalysis by $Ni(II)(acac)_2$, activated with additives of 18K6 [9]. Coordination of 18K6 with iron complexes may result in catalytically active complexes $mFe(III)(acac)_3$ n18C6 that would be transformed into $(Fe(II)(acac)_2)_p$ $(18C6)_q$ in the course of the oxidation of ethylbenzene and coordinated 18C6 would promote the oxidative transformation of Fe(II) complex into catalytically active particles according to Scheme 1 through electrophilic of O_2 connection to γ-C-atom of $acac^-$ ligand following the dioxetan formation (Scheme 1). For example, the ability of crown-ethers to catalyze electrophilic reactions of connection to γ-C-atom of $acac^-$-ligand is known [15, 16]. Steric factors appearing under coordination of 18K6 may hinder transformation of catalytically active hetero-ligand complex into inactive particles (see below).

In this article the method of control of the oxidation of ethylbenzene catalyzed by $Fe(III)(acac)_3$ into PEH with the use of 18C6 as exo ligand-modifier L^2 is examined.

2. Experimental

The oxidation of ethylbenzene (RH) with dioxygen was studied at 80°C in glass bubbling-type reactor in the presence of $Fe(III)(acac)_3$ and additives 18C6. The selectivity S_{PEH} and conversion degree C of the oxidation of ethylbenzene into PEH were determined using the formulas: $S_{PEH} = [PEH] / \Delta[RH]\cdot 100\%$ and $C = \Delta[RH] / [RH]_0\cdot 100\%$. The effects of electron-donor exoligands on w, S_{PEH} and C of the oxidation of ethylbenzene catalyzed by $Fe(III)(acac)_3$ were studied at $[Cat] = 5\cdot 10^{-3}$ mol/l and T=80°C. In this case $[PEH] = [PEH]^{max}$ and the kinetic changes in ethylbenzene oxidation under this conditions is not observed. Besides that at these conditions one can observe the maximum established effect of the influence electron-donor ligands HMPA and R_4NBr on the S_{PEH} and C of ethylbenzene oxidation at catalysis by $Fe(III)(acac)_3$ [11]. In the case of HMPA the increase in S_{PEH} from 42 (catalysis $Fe(III)(acac)_3$ without additives of L^2) to ~57% (80°C) and in conversion degree C from 4 up to 14 - 15% at $S_{PEH} = 42\%$ [12], and in the presence of CTAB ($n-C_{16}H_{33}Me_3NBr$) as exo ligand the increase in S_{PEH}^{max} from 42 to ~65%, and in conversion degree C from 4 up to 8% at $S_{PEH} = 42\%$ occurs [13].

The 18C6 concentrations were equal to $(0.5 - 5)\cdot 10^{-3}$ mol/l. It was impossible to use higher 18C6 concentrations because of the bad 18C6 solubility in the ethylbenzene.

Analysis of Oxidation Products

α-Phenylethylhydroperoxide was analyzed by iodometry. By-products, including methylphenylcarbinol (MPC), acetophenone (AP), and phenol (Ph), as well as the RH content in the ethylbenzene oxidation were examined by GLC [11]. The overall rate of the process was determined from the rate of accumulation of the all products of the oxidation of RH. A correlation between RH consumption and product accumulation was established: Δ[RH] = [PEH] + [P] + [Ph], where P = AP + MPC.

The overall rate of the oxidation of RH (**w**) and the rates of product accumulation (w_p) were determined with accuracy of ± 0.5 - 5% by a method described in [17]. The catalytic oxidation of ethylbenzene with dioxygen was carried out in the O_2 – solution two phase systems under kinetic control.

The order in which PEH, AP, and MPC formed was determined from the time dependence of product accumulation rate rations at t → 0. The variation of these rations with time was evaluated by graphic differentiation [17].

3. Results and Discussion

In reaction of the oxidation of ethylbenzene with dioxygen in the presence of catalytic system {Fe(III)(acac)$_3$(5·10^{-3} M)+18K6} (80°C) the dependence of S_{PEH} от C has extremum, as in the case of use of additives of ligands DMFA or R_4NBr. S_{PEH}^{max} = 70% ([18K6]^0 = 0.5·10^{-3} mol/l) and S_{PEH}^{max} = 75.7% ([18K6]^0 = 5·10^{-3} mol/l) in the process are higher than S_{PEH}^{max} – 65% in the case of use of CTAB as exo ligand-modifier [13] (Figure 1).

The addition of 18C6 in the oxidation of ethylbenzene with dioxygen catalyzed by Fe(III)(acac)$_3$ result in the redistribution of the major oxidation products. The significant increase in [PEH]^{max} is observed ~ 1.6 or 1.7 times at [18C6] = 5·10^{-4} mol/l, 5·10^{-3} mol/l, at that decrease in [AP] and [MPC] ~ 4, 5 times accordingly occurs. Additives of 18C6 lead to significant hindering of heterolysis of PEH with of the formation of phenol responsible for decrease in selectivity.

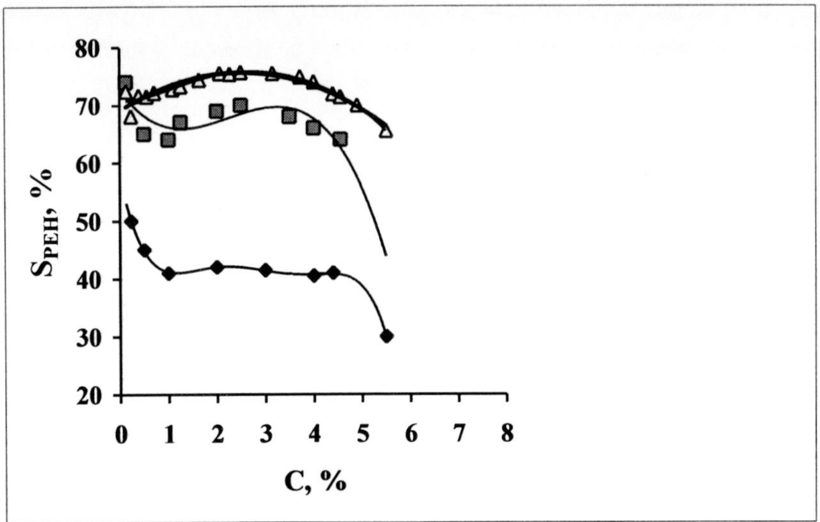

Figure 1. Dependence of SPEH от C in the reactions of the oxidation of ethylbenzene catalyzed by Fe(III)(acac)3 (♦) or catalytic systems {Fe(III)(acac)3+18C6} (Δ, □). [Fe(III)(acac)3]=5·10-3 mol/l, Δ – [18C6] = 5·10-3 mol/l, □ - 5·10-4 mol/l. 800C.

After the auto acceleration period (connected with the transformation of Fe(III) complexes into Fe(II) complexes) the w^0 rate of the accumulation of PEH in dependence of [18C6] or does not change in fact ([18C6] = $5\cdot10^{-4}$ mol/l) or decreases ~ in 2 times ([18C6]= $5\cdot10^{-3}$ mol/l). The decrease in w^0 rates of AP, MPC accumulation ~ 1.5 times ([18C6]= $5\cdot10^{-4}$ mol/l) or ~ 3.6 times ([18C6]= $5\cdot10^{-3}$ mol/l) takes place (Figures 2, 3, Table 1). As one can see the increase in $S_{PEH}{}^0$ at the initial stages of the process is connected mainly with significant decrease in accumulation rates of two major oxidation products AP and MPC. In the developed process as compared with the initial stage w_{PEH} does not change, the decrease in w_{AP+MPC} ~ 2.26 times occurs ([18C6] = $5\cdot10^{-4}$ mol/l), the w_{AP+MPC} does not change in fact and the increase in w_{PEH} ~ 2 times is observed ([18C6]= $5\cdot10^{-3}$ mol/l) (Figures 2, 3, Table 1). The major products PEH, AP, MPC are formed by parallel reactions: $w_P/w_{PEH} \neq 0$ at $t \to 0$; AP and MPC – also parallel: $w_{AP}/w_{MPC} \neq 0$ at $t \to 0$, - throughout the ethylbenzene oxidation process. (Figures 2, 3). The increase in conversion degree from C ~ 4 up to ~ 8% (at S_{PEH}=40 - (70-75.7) %) (Figure 1) takes place.

For estimation of catalytic activity of iron complexes as selective catalysts of the oxidation of ethylbenzene with dioxygen into α-phenylethylhydroperoxide we proposed to use parameter $\tilde{S}\cdot C$ by analogy with catalysis by Ni(II) or Co(II) complexes [8, 17]. $\tilde{S}$ is averaged selectivity of the oxidation into PEH characterizing change of S in the course of oxidation from S^0 at the beginning of reaction to some S^{lim}, conditional value, chosen as a standard. The value of selectivity was carried out in the limits $S^0 \leq S \geq S^{lim}$ ($S^{max} > 80\%$), C – conversion degree at S= S^{lim} [8, 17]. For comparable by value of S systems in given case the value S^{lim} =40%, approximately equal to S_{PEH} in developed reaction of the oxidation of ethylbenzene catalyzed by $Fe(III)(acac)_3$ (80^0C) was selected; C – conversion degree at which $S_{PEH} \leq 40\%$ [11,12].

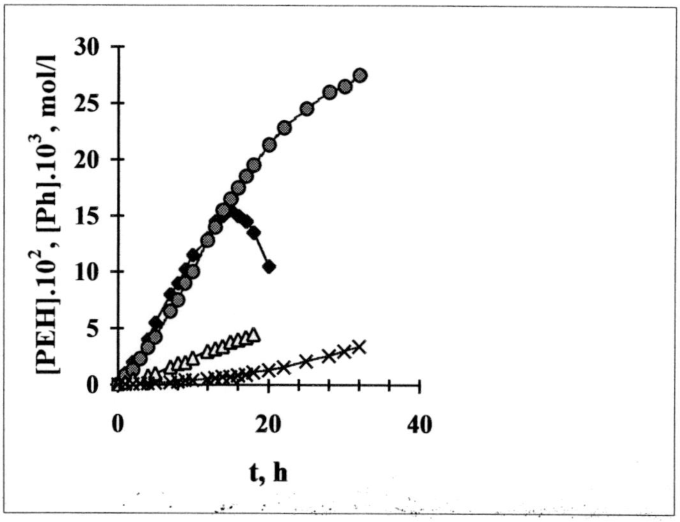

Figure 2. Kinetics of accumulation of PEH (♦,○) and Ph (Δ, x) in the reactions of the oxidation of ethylbenzene catalyzed $Fe(acac)_3$ (♦, Δ) and {$Fe(acac)_3$+18C6} (○, x) [$Fe(acac)_3$]=[18K6]=$5\cdot10^{-3}$ mol/l. 80^0C.

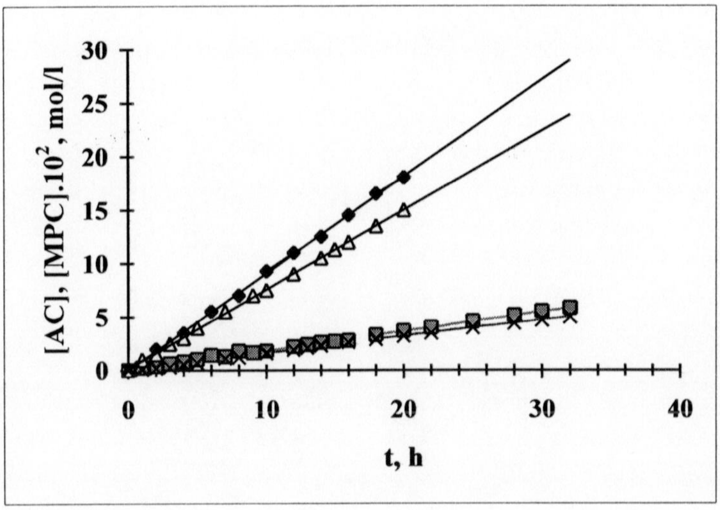

Figure 3. Kinetics of accumulation of AC (♦, □) и MPC (Δ, ×) in the reactions of the oxidation of ethylbenzene catalyzed $Fe(acac)_3$ (♦, Δ) and {$Fe(acac)_3$+18C6} (□, ×). [$Fe(acac)_3$]=[18K6]=$5\cdot10^{-3}$ mol/l. 80^0C.

Table 1. Rates of accumulation of products (mol l^{-1} s^{-1}) of the oxidation of ethylbenzene at the beginning of reaction (w^0), and in the course of the oxidation of ethylbenzene (w). Catalysis by Fe(III)$(acac)_3$ and {Fe(III)$(acac)_3$+18C6} [Fe(III)$(acac)_3$] = $5\cdot10^{-3}$ mol/l. 80^0 C

Cat	$w_{PEH}^0\cdot10^6$	$w_{AC+MPC}^0\cdot10^6$	$w_{PEH}\cdot10^6$	$w_{AC+MPC}\cdot10^6$
Fe(III)$(acac)_3$	—	—	2.90 (w_{PEH}^0=w_{PEH})	3.40 (w_P^0=w_P)
Fe(III)$(acac)_3$+18K6 ($5\cdot10^{-4}$, mol/l)	3.05	2.19	2.77	0.97
Fe(III)$(acac)_3$+18K6 ($5\cdot10^{-3}$, mol/l)	1.68	0.95	3.4	0.95

In the presence of catalytic system {Fe(III)$(acac)_3$+18K6} synergetic effect of increase in $\tilde{S}\cdot C$ parameter ~ 2,5 and 2,8 times at $[18K6]^0$=$0,5\cdot10^{-3}$ mol/l and $18K6]^0$=$5\cdot10^{-3}$ mol/l correspondingly is observed in comparison with catalysis by Fe(III)$(acac)_3$ ($\tilde{S}\cdot C$ =$2,1\cdot10^2$ (%,%)) [11]. Obtained kinetic regularities of the oxidation of ethylbenzene testify to formation presumably of $(Fe(II)(acac)_2)_p\cdot(18K6)_q$ complexes and products of their transformation in the course of oxidation (Figures 3, 4, 5 Table 1) [18]. The observing kinetic regularities are not connected with influence of 18C6. We established that in the presence of one 18C6 without iron complex auto-catalytic developing of process was observed whose initial rates were a factor of 10 lower than in case of catalysis by iron complexes, the selectivity S_{PEH} ~80% at the beginning of the reaction (C~1%), was sharply reduced with the increase in the conversion degree of ethylbenzene. At that the high rate of PhOH formation from the beginning of the reaction was observed.

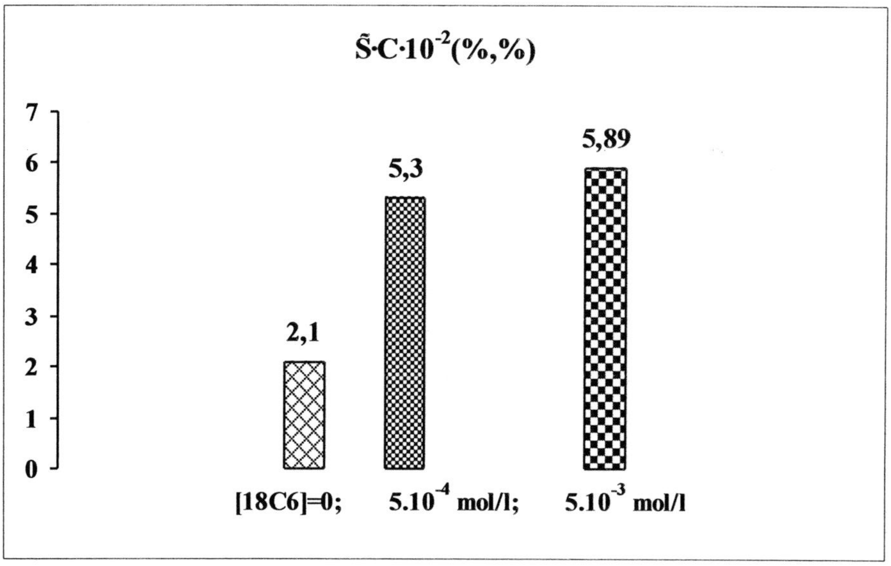

Figure 4. The values of parameter $\tilde{S}\cdot C\cdot 10^{-2}$ (%,%) in the reactions of the oxidation of ethylbenzene catalyzed by $Fe(III)(acac)_3$ or catalytic systems $\{Fe(III)(acac)_3+18C6\}$. $[Fe(III)(acac)_3]=5\cdot 10^{-3}$ mol/l, 80^0C.

As we established earlier [12] in the case of use of DMFA as electron-donor exo ligand-modifier, that activated $Fe(II)(acac)_2$ also at outer sphere coordination DMFA (hydrogen bonding) with $(acac)^-$ - ligand, the transformation of $(Fe(II)(acac)_2)_p\cdot(L^2)_q$ complexes into more active selective catalysts also took place. But in that case the value $S_{PEH}{}^{max}$ was approximately equal to S_{PEH} in the developed process of the oxidation of ethylbenzene in the presence of $Fe(III)(acac)_3$ without of L^2 additives (when Cat, $(Fe(II)(acac)_2)$ was not transformed) [11,12]. With the use of HMPA as exo ligand, that did not form H – bonds with chelate ring of $Fe(II)(acac)_2$, the transformation of $Fe(II)(acac)_2)\cdot$HMPA was not observed, although HMPA as electron-donor ligand was characterized with a higher DN value (after V. Gutmann) as compared with DMFA [11]. Supposedly, due to the favorable combination of the electronic and steric factors appeared at inner and outer sphere coordination (hydrogen bonding) of 18C6 with $Fe(II)(acac)_2$ (as also in the case of R_4NBr [13]) there is a high probability of formation of sufficiently stable hetero ligand complexes of the common structure $Fe(II)_x(acac)_y(OAc)_z(18C6)_n$, the intermediate products of the oxygenation of $(acac)^-$ ligands in the $(Fe(II)(acac)_2)_p\cdot(18C6)_q$ complexes (Schemes 2) by analogy to the catalysis by acetyl acetone dioxygenase (M = Fe(II)) (Dke 1) that results in the $S_{PEH}{}^{max}$ and C increase.

$$\{Fe(III)(acac)_3+18K6\}\rightarrow Fe(II)(acac)_p\cdot(18K6)_q+O_2\rightarrow Fe(II)_x(acac)_y(OAc)_z(18K6)_n. \qquad (I)$$

In the case of catalysis by $Ni(II)(L^1)_2$ (L^1=acac$^-$) complexes, activated with additives of electron donor L^2 ligands [3-10], the another mechanism of the oxidative cleavage of the L^1 acetylacetonate ligand under the action of O_2 was established by us [7,8]. Controlled by L^2 ligand regio-selective connection of O_2 to nucleophilic carbon γ-atom of the L^1 ligand through the break of the formed transition cycle configuration leads to formation of (OAc^-) ion, acetaldehyde, elimination of CO that results in the formation of catalytic active $Ni_x(L^1)_y(L^1{}_{ox})_z(L^2)_n$ complexes ("A"). Similar changes in complexes' ligand environment under the action of O_2 with the formation of carbonyl compounds and elimination of CO were established in the reactions of oxidative cleavage of β-diketonates, catalyzed with the only known to date Ni(II)-containing dioxygenase – acireductone dioxygenase, ARD [19], and in reactions of oxygenation imitating the action of quercetin 2,3-dioxygenase (Cu, Fe) [14, 20].

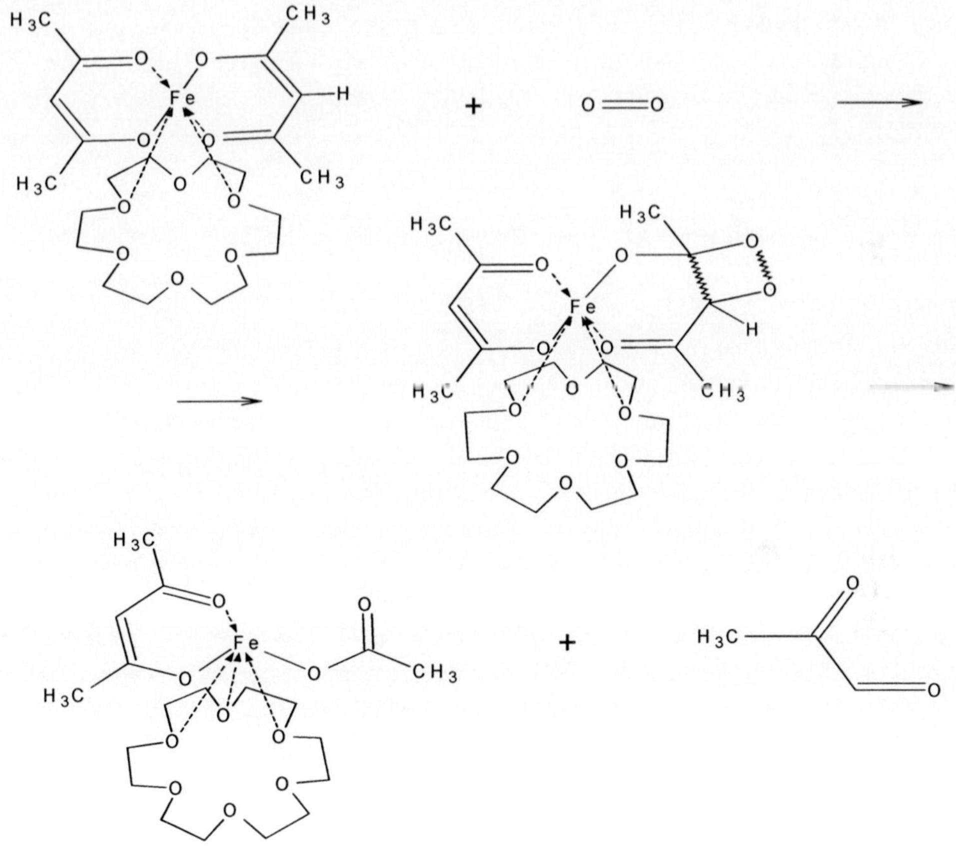

Scheme 2. The proposed scheme of oxygenation of $Fe(II)(acac)_2$ in the presence of 18C6.

We proposed the method for estimation of catalytic activity of complexes formed *in situ* at the beginning of reaction and in developed process, at micro steps of the oxidation of ethylbenzene: chain initiation (O_2 activation by Cat (metal complex)) and chain propagation ($Cat + RO_2^{\bullet} \rightarrow$) [8-13, 17] by simplified scheme assuming quadratic termination of chain and equality to zero of rate of homolytic decomposition of ROOH. In the framework of radical-chain mechanism the chain termination rate w_{term} in this case will be (1):

$$\mathbf{w_{term}} = k_6[RO_2^{\bullet}]^2 = k_6\left\{\frac{\mathbf{w}_{PEH}}{\mathbf{k}_2[RH]}\right\}^2 \qquad (1)$$

where w_{PEH} – rate of PEH accumulation, k_6 – constant of rate of quadratic chain termination reaction; k_2 – constant of rate of chain propagation reaction ($RO_2^{\bullet}$ + RH→).

We established that iron complexes $Fe(II,III)(L^I)_n$ ([Cat]=$(0.5\text{-}5)\cdot10^{-3}$ mol/l)) were inactive in PEH homolysis, products MPC and AP were formed at parallel steps of chain propagation (Cat + $RO_2^{\bullet}$→) and quadratic chain termination. Actually, $w^0 \sim [Cat]^{1/2}$ and $w_i^0 \sim$ [Cat] and linear radicals termination on catalyst may be not taken into account. In the case of quasi-stationarity by $RO_2^{\bullet}$ radicals $w_{term} = w_i$, and calculated by formula (1) w_{term} is the measure of iron(II) complexes activity in relation to molecular O_2. Discrepancy between w_{AP+MPC} and w_{term} (Fig. 6, Table 2) in the case of absence of linear chain termination is connected with additional formation of alcohol and ketone at the steps of chain propagation (Cat + $RO_2^{\bullet}$→) (2):

$$w_{pr.} = w_{AP+MPC} - w_{term} \quad (2)$$

The direct proportional dependence of w_{pr}^0 on [Cat] testifies in favor of iron(II) complexes participation at chain propagation (Cat + $RO_2^{\bullet}$→).

We suggest that the conditions $w^0 \sim [Cat]^{1/2}$ and $w_i^0 \sim$ [Cat] will be fulfilled also in the presence of additives of 18C6.

Except theoretical consideration in [21], activity of transition metals complexes $M(L^1)_n$ (M=Ni, Co, Fe, L^1 = $acac^-$, $enamac^-$) at reaction of chain propagation (Cat + $RO_2^{\bullet}$→) of the oxidation of ethylbenzene is estimated only in our works [8-13, 17].

Investigation of reaction ability of peroxide complexes [LM-OOR] (M=Co, Fe) preliminary synthesized by reactions of compounds of Co and Fe with ROOH or $RO_2^{\bullet}$-radicals [22-24] confirms their participation as intermediates in reactions of the oxidation of hydrocarbons. Obviously, the schemes of radical-chain oxidation including intermediate formation of [LM-OOR] [22-25] with further homolytic decomposition of peroxo-complexes ([LM-OOR]→R′C=O (ROH) + $R^{\bullet}$) may explain parallel formation of alcohol and ketone under the oxidation of ethylbenzene in the presence of $M(L^1)_n$ (M=Fe(II), L^1 = $acac^-$) and their complexes with 18C6.

The chain initiation in the oxidation of ethylbenzene with dioxygen in the presence of $Fe(III)(acac)_3$ or {$Fe(III)(acac)_3$+18C6} can be represented by the following reaction:

$$Fe(III)(acac)_3\ ((Fe(III)(acac)_2)_m\cdot(18C6)_n) + RH \rightarrow$$
$$\rightarrow Fe(II)(acac)_2\ ((Fe(II)(acac)_2)_p\cdot(18C6)_q) \ldots Hacac + R^{\bullet} \quad (II)$$

The reaction (II) and interaction of the resulting Fe(II) complex with dioxygen appear to be responsible for chain initiation in the reaction catalyzed by $Fe(III)(acac)_3$ or {$Fe(III)(acac)_3$+ 18C6}.

As one can see from the Table 2 the increase in S_{PEH}^0 from ~ 40 - 50 to ~ 65 - 70% at the initial stages of the oxidation of ethylbenzene catalyzed by system {$Fe(III)(acac)_3$+18K6} ($[18K6]^0$=$0.5\cdot10^{-3}$ mol/l, $5.0\cdot10^{-3}$ mol/l), at catalysis by $(Fe(II)(acac)_2)_x\cdot(18C6)_y$ complexes, is connected mainly with decrease in the rates of the formation of AP and MPC at micro steps of chain propagation (Cat + $RO_2^{\bullet}$→). The decreases in w_{pr}^0 1.6 and 3.5 times correspondingly take place. At that w_i^0 is either constant ([18K6] = $0.5\cdot10^{-3}$ M), or is reduced in 3.3 times ([18K6] = $5\cdot10^{-3}$ M). The drop of w_{pr}^0 and w_i^0 occurs evidently due to electron and steric factors, appeared at the inner and outer sphere coordination of 18C6 with Fe(II)(acac)$_2$, that may decrease the probability of the coordination of $RO_2^{\bullet}$ and O_2 with the metal ion and formation primary complexes with O_2 and / or $RO_2^{\bullet}$ following the formation of the active particles of superoxide and peroxide types.

Table 2. Rates of accumulation of ethylbenzene oxidation products (mol l^{-1} s^{-1}) at the beginning of reaction (w^0), and calculated rates of chain initiation (w_i^0) and chain propagation (w_{pr}^0) and ($w_i^0/w_{пр}^0$)·100%. Cat = Fe(III)(acac)$_3$ and {Fe(III)(acac)$_3$+18C6}. [Fe(III)(acac)$_3$]=5·10^{-3} mol/l. 80^0 C

Cat	$w_{PEH}^0\cdot10^6$	$w_{AC+MPC}^0\cdot10^6$	$w_i^0\cdot10^7$	$w_{пр}^0\cdot10^6$	$(w_i^0/w_{пр}^0)100\%$
Fe(III)(acac)$_3$	2.90	3.40	0.79	3.32	2.38
{Fe(III)(acac)$_3$+18C6 (5·10^{-4}, mol/l)}	3.05	2.19	0.80	2.11	3.79
{Fe(III)(acac)$_3$+18C6 (5·10^{-3}, mol/l)}	1.68	0.95	0.24	0.93	2.58

The ratio $w_i^0/ w_{pr}^0 \approx 2.4-3.8\%$ means that the (Fe(II)(acac)$_2$)$_x$ (18K6)$_y$ complexes are more active in chain propagation (Cat + $RO_2^{\bullet}\rightarrow$) than in chain initiation as in case of catalysis by systems {Fe(III)(acac)$_3$ + L^2} (L^2= HMPA, DMFA, R_4NBr) [12, 13]. AP and MPC form mainly in chain propagation (through the [Fe(III)–OOR] intermediate complex [19–22].

Furthermore, this value points to an essential contribution from the chain propagation to the overall rate of the process and mechanism of catalysis. At catalysis by Ni(II) complexes the ratio $w_i^0/ w_{pr}^0 \approx$ 11-50% and is sensitive to the nature of the ligand environment of nickel ion in Ni(II)(L^1)$_2$ (L^1 = acac$^-$, enamac$^-$) complexes. In the presence of 18C6 at the catalysis by Ni(II)(L^1)$_2$ (18K6)$_n$ complexes w_i^0 increases significantly - in 5 (chelate group (O/NH)) or 17 times (chelate group (O/O)), at that w_{pr} decreases sharply, and w_{pr}^0 = 0 (Ni(II)(L^1)$_2$ (18K6)$_n$ (n=1,2, L^1=acac$^-$; (n=1, L^1= enamac$^-$) [23].

The Fe(II)$_x$(acac)$_y$(OAc)$_z$(18C6)$_n$ complexes formed in the process of the oxidation of ethylbenzene catalyzed by system {Fe(III)(acac)$_3$+18C6} are likely inactive in the heterolytic decomposition of PEH to PhOH and acetaldehyde and in the reaction with the $RO_2^{\bullet}$ radicals (the chain propagation (Cat + $RO_2^{\bullet}\rightarrow$)) (Figures 2, 3).

4. CONCLUSION

Thus, the use of additives of 18C6 in the oxidation of ethylbenzene with dioxygen catalyzed by Fe(III)(acac)$_3$ allowed the increase of values of selectivity, S_{PEH}, conversion degree, C, and PEH yield that exceed analogous parameters in the presence of Fe(III)(acac)$_3$ without 18C6. $\tilde{S}\cdot C$ parameter grows 2.5 and 2.8 times (synergetic effect). The dependence of S_{PEH} on C has the extremum S_{PEH}^{max} that by ~ 34% more than S_{PEH} at catalysis by Fe(III)(acac)$_3$ without 18C6 at the same C values. The kinetic regularities testify to the formation of catalytic active (Fe(II)(acac)$_2$)$_p$·(18C6)$_q$ complexes and products of their transformation.

The increase in S_{PEH} is accompanied by the significant increase in the PEH concentration and hindering of the major products MPC and AP and also PhOH formation rates. As in the absence of 18C6 additives MPC and AP are formed at the parallel to PEH formation steps, MPC and AP – also at parallel steps in the course of the investigated process.

The method of evaluation of the activity of transition metal complexes in the micro steps of the chain-radical oxidation of ethylbenzene proposed by us earlier has been applied for the constructed catalytic systems on the basis of iron complexes and 18C6. It was established that the rate and selectivity of the oxidation of ethylbenzene into PEH at the catalysis by $(Fe(II)(acac)_2)_p\cdot(18C6)_q$ complexes, formed at the initial stages of the process, depend on the ratio between rates of chain initiation (w_i, activation O_2) and chain propagation (w_{pr}, Cat+$RO_2^{\bullet}\rightarrow$). The increase in $S_{PEH}{}^0$ from ~40 - 50 to ~ 65 - 70% is connected mainly with a decrease in catalytic activity of $(Fe(II)(acac)_2)_p\cdot(18C6)_q$ complexes at reaction of chain propagation (Cat + $RO_2^{\bullet}\rightarrow$) (through the [Fe(III)–OOR] intermediate complex). The ratio $w_i^0 / w_{pr}^0 \approx 2.4 - 3.8\%$ means that the iron complexes $(Fe(II)(acac)_2)_p$ $(18K6)_q$ are more active in chain propagation (Cat + $RO_2^{\bullet}\rightarrow$) than in chain initiation step as in the case of catalysis by the systems {$Fe(III)(acac)_3 + L^2$} (L^2=R_4NBr, HMPA, DMFA) [12,13].

The active selective $Fe(II)_x(acac)_y(OAc)_z(18C6)_n$ catalysts formed in the course of the oxidation of ethylbenzene are likely also to be inactive in heterolytic PEH decomposition and in the reaction with $RO_2^{\bullet}$ radicals. The 'dioxygenase-like' mechanism of the dioxygenation of the $(Fe(II)(L^1)_2)_p(R_4NBr)_q$ complexes into new selective catalysts in the process of the oxidation of ethylbenzene is suggested.

References

[1] K. Weissermel, H.-J. Arpe, *Industrial Organic Chemistry*, 3nd ed., transl. by C.R. Lindley. New York, VCH, 1997, 427 pp.

[2] D. Mansuy, In: *The Activation of Dioxygen and Homogeneous Catalytic Oxidation. /* Eds. Barton D.H.R., Martell A.E., Sawyer D.T. New York, Plenum Press, 1993, 488 pp.

[3] L.A. Mosolova, L.I. Matienko, Z.K. Maizus, Izv. AN SSSR, Ser. Khim., N 2, pp. 278-281, 1980 (in Russian).

[4] L.A. Mosolova, L.I. Matienko, *Neftekhimiya*, Vol. 25, N 4, pp. 540-545, 1985 (in Russian).

[5] L.A. Mosolova, L.I. Matienko, I.P. Skibida, *Kinetika i kataliz*, Vol. 29, N 5, pp. 1078-1083, 1988 (in Russian).

[6] L.A. Mosolova, L.I. Matienko, Z.K. Maizus, Izv. AN SSSR, *Ser. Khim.*, N 4, pp. 731-735, 1981 (in Russian).

[7] L.A. Mosolova, L.I. Matienko, I.P. Skibida, *Kinetika i kataliz*, Vol. 28, N 2, pp. 479-484, 1987 (in Russian).

[8] L.I. Matienko, L.A. Mosolova, Izv. AN, *Ser. Khim.*, N 1, pp.55-60, 1999 (in Russian).

[9] L.A. Mosolova, L.I. Matienko, I.P. Skibida, Izv. AN, *Ser. Khim.*, N 8, pp.1406-1410, 1994 (in Russian).

[10] L.A. Mosolova, L.I. Matienko, I.P. Skibida, Izv. AN, *Ser. Khim.*, N 8, pp.1412-1416, 1994 (in Russian).

[11] L.I. Matienko, L.A. Mosolova, *Kinetika i kataliz*, Vol. 46, N 3, pp. 354-360, 2005 (in Russian).
[12] L.I. Matienko, L.A. Mosolova, *Neftekhimiya*, 2006 (in prints).
[13] L.I. Matienko, L.A. Mosolova, In: *New Aspects of Biochemistry Physics. Pure and applied sciences* / Ed. by S.D.Varfolomeev E.B. Burlakova, A.A. Popov, G.E. Zaikov. New York: Nova Sience Publ., 2006 (in prints).
[14] G. D. Straganz, B. Nidetzky, *J. Am. Chem. Soc.*, Vol. 127. N 35. pp. 12306-12314, 2005.
[15] H. - J.Schneider, R. Busch, R. Kramer, e.a., In: Nucleophility. Adv. Chem. Ser. / Eds. Harris J.M., McManus S.P., Washington, *Am. Chem. Soc.*, 1987, 482 pp.
[16] E.V. Demlov, Izv. AN SSSR, *Ser. Khim.*, N 11, pp. 2094-2101, 1995 (in Russian).
[17] L.I. Matienko, L.A. Mosolova, Izv. AN SSSR, *Ser. Khim.*, N 4, pp. 689-693, 1997 (in Russian).
[18] V.A. Golodov, Ross. *Khim. Zh.*, Vol. 44, N 3, pp 45-57, 2000 (in Russian).
[19] Y. Dai, Th. C. Pochapsky, R.H. Abeles, *Biochem.*, Vol. 40. N 21. pp. 6379-6387, 2001.
[20] M. Utaka, M. Hojo, Y. Fujiu, A. Takeda, *Chem. Lett.*, N 6, pp. 635-638, 1991.
[21] N.M. Emanuel, D.Gal, Ethylbenzene oxidation. Model reaction, Moscow, *Nauka*, 1984, 375 pp. (in Russian).
[22] R.D.Arasasingham, Ch.R. Cornman, A.L. Balch, *J. Am. Chem. Soc.*, Vol. 111, N 20, pp. 7800-7805, 1989.
[23] F.A. Chaves, J.M. Rowland, M.M. Olmstead, P.K. Mascharak, *J. Am. Chem. Soc.*, Vol. 120, N 35, pp. 9015-9027, 1998.
[24] E. Solomon-Rapaport, A. Masarwa, H. Cohen, D. Meyerstein, *Inorg. Chim. Acta*, N 299, pp. 41-46, 2000.
[25] A.E. Semenchenko, V.M. Solyanikov, E.T. Denisov, *Zh. Phiz. Khimii*, Vol. 47, N 5, pp. 1148-1151, 1973 (in Russian).

In: Preparation and Properties of Monomers, Polymers... ISBN 978-1-60021-557-5
Eds: A. Ballada and G. E. Zaikov, pp. 133-136

Chapter 14

ENHANCED PHOTO AND THERMAL OXIDATIVE STABILITY OF CHARGE-TRANSFER COMPLEXES OF CONJUGATED POLYMER

Sergei Arnautov*, Elena Nechvolodova*, Sergei Lomakin and Alexander Shchegolikhin*****

*N.N. Semenov Institute of Chemical Physics of the Russian Academy of Sciences, Kosygin st., 4, Moscow 119991, Russia

**N.M. Emanuel Institute of Biochemical Physics of the Russian Academy of Sciences, ul. Kosygina 4, Moscow 119991, Russia

Practical employment of conjugated polymers hinges on improving their photo and thermal stability. To clarify degradation mechanisms of charge-transfer complexes of MEH-PPV with some organic acceptors (for e.g. trinitrofluorenone (TNF)), the corresponding MEH-PPV/TNF blends have been systematically studied by various methods. In the donor/acceptor molar ratios range of 5-25 mole % these compositions have been found to exhibit a systematic considerable gain in photo and thermal oxidative stability. According to a model-based kinetic analysis data, the predicted lifetimes for MEH-PPV/TNF blends at practically important temperatures (60-100°C) are three orders of value higher than those of the neat conjugated polymer.

It is now widely recognized that thermal oxidative stability of a conjugated polymer (CP) is very important for both its processing and exploitation, while photo- and/or thermo-oxidation of the polymers is a major factor which leads to shortening lifetime and lowering efficiency of various devices based thereon [1,2]. Therefore, elucidation of mechanisms of such degradation as well as correct service life predictions are necessary for optimizing the CP-based devices.

Poly[(2-methoxy-5-(2'-ethylhexyloxy)-1,4-phenylene vinylene] (MEH-PPV) has been all-embracingly investigated for a decade and is now a "classic" model for tackling generic problems of CPs[3]. It has been reported recently [4] that MEH-PPV is capable of forming

* E-mail: saa@chph.ras.ru

intermolecular CTCs with small organic acceptors and, in particular, with 2,4,7-trinitro-fluorene-9-one (TNF).

Drop-cast films of both pristine MEH-PPV (Sigma-Aldrich, MW=125,000) and of MEH-PPV blends with TNF (Aldrich, m.p.171, purity > 98%) acceptor were prepared from chlorobenzene by varying the polymer concentration in the range of 0.5–2 g l^{-1} and the molar TNF:MEH-PPV ratio in the range of 0.01–1.

Susceptibility of neat MEH-PPV to photo oxidation and comparison of the latter with that of MEH-PPV/TNF blends is illustrated in Figure 1. An immediate blue shift of λ_{max} and progressive reduction of A_{max} were observed in the course of irradiation, and are explained by shortening the π-electron delocalization length of the polymer. FTIR spectra of the same films measured before and after irradiation have revealed new bands emerging at 1279 and 1691 cm^{-1} unambiguously indicating [5] photo oxidation of PPV. It is quite evident that MEH-PPV/TNF blends exhibit markedly improved photo-oxidative stability as compared to that of the neat polymer.

Moreover, for TNF contents increasing in the range of 5-25 mole %, the blends demonstrated a clear systematic gain in their thermal-oxidative stability. Of importance is the fact that both components of the blends have revealed a synergistic behaviour in terms of the stability enhancement (Figure 2). Comparative experiments have shown that fullerene C60 is also capable of stabilizing MEH-PPV, but when present in much higher concentration.

Experimental data obtained by other methods indicate to spectacular non-covalent interactions between the donor and the acceptor within the same (5-25 mole %) range of TNF contents. Partial electron density transfer from MEH-PPV polymer onto TNF molecules involved in the corresponding CTC (as it follows, e. g., from the FTIR and Raman data), should in principle reduce the susceptibility of the polymer to oxygen.

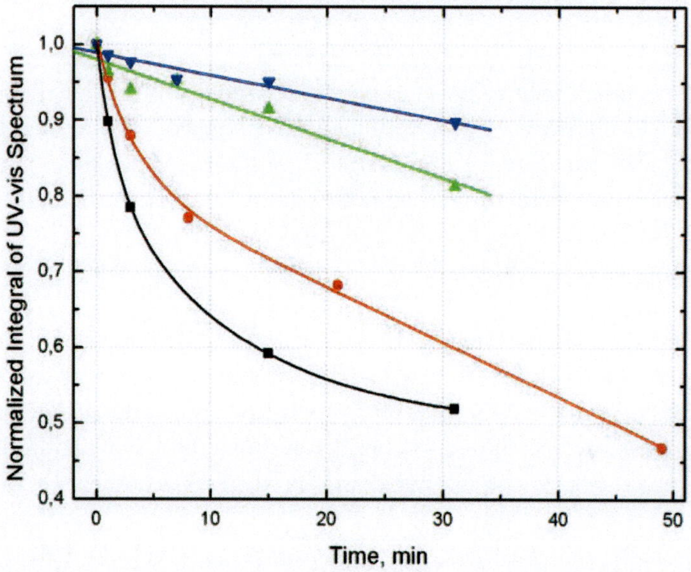

Figure 1. The integral UV-VIS absorption *vs.* irradiation time dependences. From bottom to top (mole/mole): neat MEH-PPV (black); MEH-PPV:TNF=1:0.3 (red); MEH-PPV:TNF=1:0.6 (green); MEH-PPV:TNF=1:1 (blue). Lines are given as guidance for eyes.

It is quite probable that quenching of the triplet states excited on the macromolecules can effectively prevent a triplet-triplet annihilation reactions which are normally feasible in the presence of reactive singlet oxygen [6].

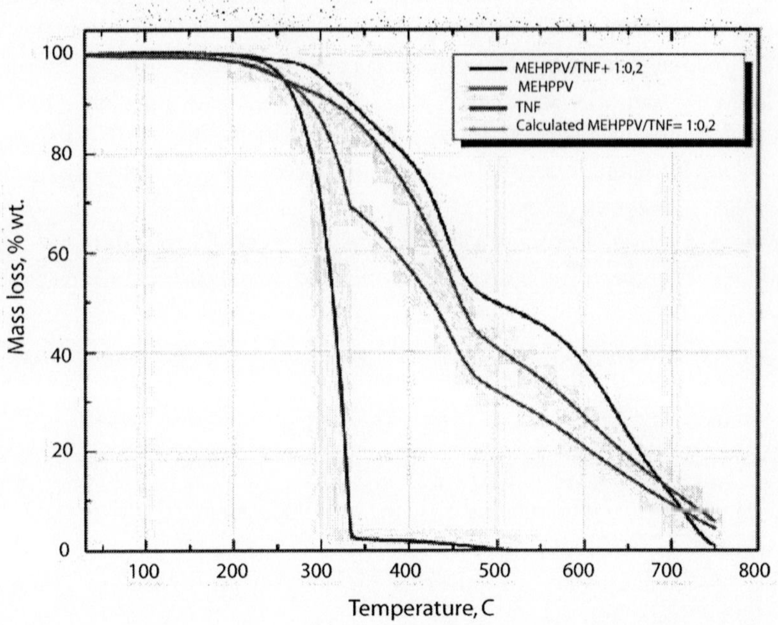

Figure 2. Experimental TGA curves for neat TNF (blue), neat MEH-PPV (red), MEH-PPV:TNF = 1:0.2 (mol.) blend. A simulated TGA curve built by simple addition of the components partial contributions for the same MEH-PPV/TNF blend is shown in green.

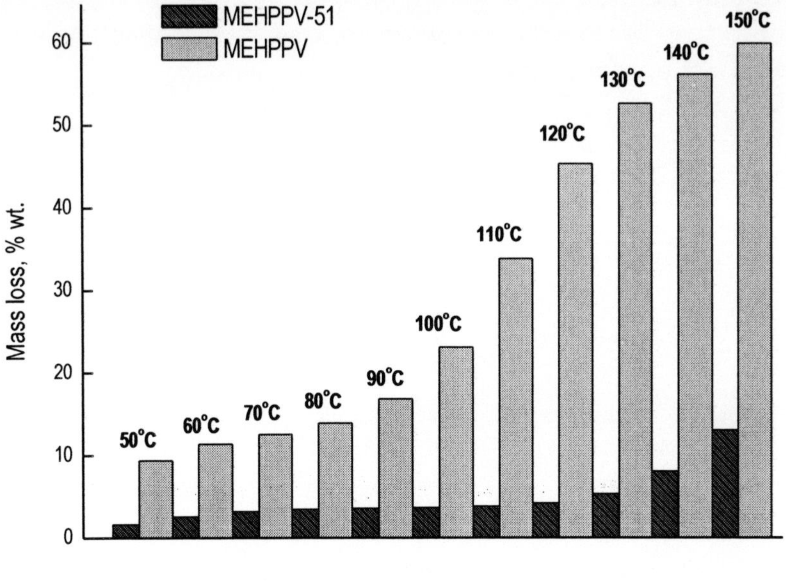

Figure 3. Comparison of one year predicted life-times behaviours for the neat MEH-PPV and MEH-PPV/TNF = 1/0.2 (mol.) blend in ambient (air) atmosphere.

Kinetic study undertaken for both MEH-PPV/TNF blends and the pristine polymer (using a model-free Friedman analysis, then nonlinear regression analysis of the kinetic parameters [7], then life-time prediction) has indicated that a four-step thermal-oxidative degradation processes are operative in the system. The experimental results obtained in this study infer that the complexed TNF (or corresponding TNF/MEH-PPV charge-transfer complex) is capable of effectively intercepting radical species that could trigger thermal oxidation processes in the system. The same study has provided also extra evidence to the stabilizing effect of the acceptor in the blends: the lifetimes predicted for MEH-PPV/TNF blends at practically important temperatures (60-100oC) are three orders of value higher than those of the neat conjugated polymer (Figure 3).

Resuming, we present here a practically feasible approach to improving the photo- and thermal-oxidative stability of conjugated polymers by virtue of CTC formation thereof with organic acceptors.

References

[1] A. T. H. Koch, D. Beljonne, N. T. Harrison, J. L. Bredas, N. Haylett, R. Daik, W. J. Feast, and R. H. Friend, *Optical Materials*, 1998, 8, 145.

[2] F. Fitrilawati, M. O. Tjia, S. Pfeiffer, H. H. Horhold, A. Deutesfeld, H. Eichner, and C. Bubeck, *Optical Materials*, 2003, 21, 511.

[3] M. Atreya, S. Li, E.T. Kang, K.G. Neoh, Z.H. Ma, K.L. Tan, and W. Huang, *Polymer Degradation and Stability*, 1999, 65, 287.

[4] A.A. Bakulin, S.G. Elizarov, A.N. Khodarev, D.S. Martyanov, I.V. Golovnin, D.Yu. Paraschuk, M.M. Triebel, I.V. Tolstov, E.L. Frankevich, S.A. Arnautov, E.M. Nechvolodova, *Synthetic Metals*, 2004, 147, 221.

[5] L.J. Rothberg, M. Yan, F. Papadimitrakopoulos, M.E. Galvin, E.W. Kwock, T.M. Miller, *Synthetic Metals*, 1996, 80, 41.

[6] T.Q. Nguyen, I.B. Martini, J.Liu, and B. J. Schwartz, *Journal of Physical Chemistry* B, 2000, 104,237.

[7] J. Opfermann, *J. Thermal Anal. Cal*, 2000, 60, 641.

In: Preparation and Properties of Monomers, Polymers… ISBN 978-1-60021-557-5
Eds: A. Ballada and G. E. Zaikov, pp. 137-143

Chapter 15

PREPARATION AND INVESTIGATION OF PHYSICAL AND CHEMICAL PROPERTIES OF IONIC MAGNETIC FLUIDS ON THE BASIS OF COBALT FERRITE

F. S. Bayburtskiy**[*1]**, Yu. V. Baldokhin**[1]**, A. S. Solomatin**[1]**,
A. A. Novakova**[2]**, L. A. Kulikov**[2]**, N. I. Ivanova**[2]**,
***A. V. Volkov**[2] **and T. V. Tsyganova**[2]*

[1]N. M. Emanuel Institute of Biochemical Physics, Russian Academy of Sciences
4, Kosygin str., Moscow 119991, Russian Federation
[2]V. Lomonosov Moscow State University, Department of Chemistry,
Vorobyoby gory, Moscow, 119899 Russian Federation

ABSTRACT

Size-controlled synthesis of cobalt-ferrite nanoparticles, their passivation and peptization as stable ferrofluids are reported. Transmission electron microscopy and Mossbauer spectroscopy were used as characterization techniques. Particles with little change in size distribution, in the 10 – 15 nm diameter ranges, were obtained using stirring speeds between 2700 and 8100 rpm. The anomalous diffusion has been used to explain the nanoparticle size-control mechanism.

Keywords: Magnetic fluids; cobalt ferrite; transmission electron microscopy; size control; Mossbauer spectroscopy; superparamagnetism; nanoparticle synthesis.

INTRODUCTION

The chemical synthesis of spinel ferrite-based nanoparticles and their surface passivation and stabilization as ionic magnetic fluids (IMF) may represent a very important step towards

* E-mail: Chembio@sky.chph.ras.ru, Bayburt@mail.ru

the engineering of specific magnetic carriers. In particular, the capability to synthesize size-controlled cobalt-ferrite based nanoparticles plus the possibility of using radioactive ^{60}Co to produce enriched IMF opens new perspectives, as for instance in the targeting of cancer cells using antibody-coated nanoparticles. Size control of spinel ferrite-based nanoparticles may be achieved by controlling the physical parameters and the chemical conditions under which the chemical condensation reaction occurs [1, 2]. Furthermore, surface cation exchange may be successfully used to passivate the nanoparticle surface, thus preventing chemical transformations [3]. Finally, the control of the pH, ionic strength, and counterion nature provide the necessary conditions for the IMF stability [4]. Transmission electron microscopy (TEM) has been used as the most traditional technique for nanoparticle sizing of IMF samples [5]. Besides the size characterization provided by the TEM data, Mossbauer spectroscopy can be used as a powerful technique to characterize several aspects, specially in iron-based materials, such as chemical composition, crystal structure, and magnetic behavior [6].

In this study, we report the chemical synthesis of size-controlled cobalt-ferrite nanoparticles, the passivation procedure, the peptization of the nanoparticles as highly stable ionic magnetic fluids, and the material characterization using TEM and Mossbauer spectroscopy. The microscopy data provide information concerning the particle-size polydispersity, which will be discussed in terms of the synthesis conditions. Mossbauer spectroscopy, on the other hand, not only indicates the magnetic phase obtained but also revels the magnetic ordering associated with the nanoparticles.

MATERIALS AND METHODS

Stable cobalt-ferrite-based ($CoFe_2O_4$) IMF were obtained after performing a three-step procedure. First, cobalt-ferrite nanoparticles were synthesized by coprecipitation Co (II) and Fe (III) in alkalin medium.

Stabilization of magnetic particles by solutions of electrolytes carried out three variants.

Variants I. The starting solution containing 6 ml of concentrated nitric acid, 250 ml of aqueous solution of cobalt nitrate (0.005 mol / l), and 250 ml of aqueous solution of iron chloride (0,010 mol / l) was heated to 95°C, under vigorous stirring. Aqueous solution of sodium hydroxide (1.0 mol / l) was then quickly added to the stirring solution until pH 12 was reached. About 120 ml of the sodium hydroxide solution was required to bring the pH to about 12. After removing the aqueous solution, nitric acid (2.0 mol / l) was added to the precipitate until pH 3 was reached. Then, the precipitate rested overnight on top of a magnet. Second, passivation of the nanoparticles surface was achieved by Fe (III) enrichment, after treatment with ferric nitrate. Third, peptization of the nanoparticle was obtained by controlling the surface charge density while reducing the ionic strength of the aqueous medium. An aqueous triethylamine (0.1 mol / l) solution was added drop wise to the acid precipitate until pH 6 was reached. Three different samples were prepared using different stirring conditions, namely, S13, S23, and S33, respectively, at 2700, 5400, and 8100 rotation per minute (rpm). Samples S13, S23 and S33 were treated with 10 ml of hot ferric nitrate (0.5 mol / l) under stirring, during about 3 min. The precipitate was then washed through several cycles with acetone and deionized water to reduce the ionic strength of the medium. Finally, perchloric acid ($HClO_4$) was added to the precipitate until pH 2 was reached at which point peptization occurred.

Variants II. The so-obtained cobalt-ferrite ($CoFe_2O_4$) is then acidified, dispersed into water, giving an ionic acidic magnetic fluid. The nanoparticles are positively charged with nitrate (NO_3^-) counter-ions. This magnetic fluid is then taken as a precursor for the samples studied. Adsorption of an ionic organic ligand (trisodium citrate) on these positive particles leads to an ionic magnetic fluid, stable in an aqueous medium (pH > 3); in this case, the nanoparticle surface charge is negative and neutralized by Na^+ counter-ions.

Variants III. The nanoparticles are positively charged with nitrate (NO_3^-) counter-ions. This magnetic fluid is then taken as a precursor for the samples studied. The tartrate-based nanoparticles are obtained after treating the acid ionic magnetic fluids with aqueous solution of sodium tartrate.

Thus water-based ionic ferrofluids with different average particle diameters, depending on the coprecipitation conditions, are obtained. Here we report only on Co-ferrite based samples (powder sample and the corresponding ferrofluid). Results of the investigation of Fe–Mn–, and Ni–ferrites will be reported elsewhere.

In the present work we in details investigated the ionic magnetic fluids prepared on the first way. All of them were well diluted with distilled water in any ratio and kept for a long time the aggregation and sedimentation stability.

DISCUSSION

Control of the nanoparticle size happens during the first synthesis step by changing the stirring speed in the range of 2700 – 8100 rpm. It was found that the higher the stirring speed the smaller the average mean particle diameter. However, the standard deviation associated with the lognormal particle-size distribution did not change when the nanoparticle size increased by about 50 %, as discussed below. The understanding of the nanoparticle growth mechanism through chemical condensation reaction in solution and under vigorous stirring, has not been the focus of these investigations, despite the huge impact it may have from fundamental and applied viewpoint.

However, the decrease of the average particle size as the stirring speed increases may indicate that the nanoparticle growth process is mainly dominated by the nanoparticle diffusion throughout the medium instead of the diffusion of the aqueous cations into the nanoparticles surface. The seems to be an obvious conclusions, once cations are everywhere in the medium while the nanoparticles represent a small volume fraction. Nevertheless, the distinction between normal and anomalous diffusion of the nanoparticle may hold the key to understand the growth process. Under normal diffusion the nanoparticle growth process may keep the same growth kinetics along the mean-free path while under anomalous diffusion the growth kinetics may switch between two distinct paths, with very different growth velocities. The two distinct growth paths may occur along with a sequence of short flies or during the long flies, typical of the anomalous diffusion [7].

Anomalous diffusion of particles inside a turbulent fluid has been recently reported in the literature [8]. Under such a scenario, the nanoparticles growth is more likely to occur during a sequence of short flies, when space localization is higher. As the stirring speed increases, the time the nanoparticle spend during a sequence of short flies also increases, allowing the onset of a higher density of the cobalt-ferrite nanoparticles, from where the particle-size polydispersity profiles were taken, were obtained using the JEOL 100 CXII equipment. The solid line in figure 1 represents the best-fit curve of the particle-size histogram using the standard approach, i. e., the lognormal distribution function [9].

$$P(D) = \frac{\exp(-2\sigma^2)}{D_m \sigma \sqrt{2\pi}} \exp\{-\ln^2(D/D_m)/2\sigma^2\} \qquad (1)$$

with a most probable particle diameter of Dm = 15.1 ± 0.1 and 11.4 ± 0.1 nm and a standard deviation of σ = 0.30 ± 0.01 and 0.31 ± 0.01, for samples S13 and S33, respectively.

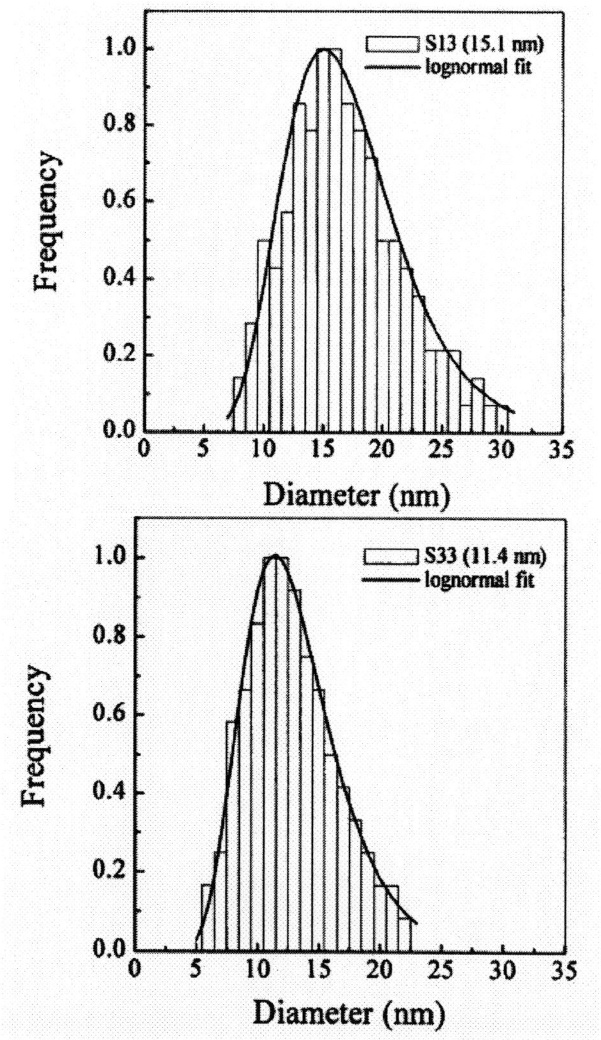

Figure 1. Typical particle size histogram obtained from the TEM data. The solid lines represent the best fit of the data according to equation (1).

Note that the lognormal distribution function described in equation (1) provides a full-width al half-maximum (ΔD) that spans from f^{-1} *Dm* to *f Dm*, where f = exp [σ (2 ln) $^{1/2}$]. The f factor also described the asymmetry of the lognormal distribution function, once $f = -(f$ *Dm* – *Dm*) / (f^{-1} *Dm* – *Dm*). Therefore, from the fitting of the TEM histogram using equation (1) one obtains the full-width al half-maximum of ΔD = 10.9 nm (σ = 0.30) and ΔD = 8.5 nm (σ = 0.31) associated to samples S13 and S33, respectively. In figure 2 (*a* and *b*) the electronic microphoto of ferrite nanoparticles in a magnetic fluid is shown.

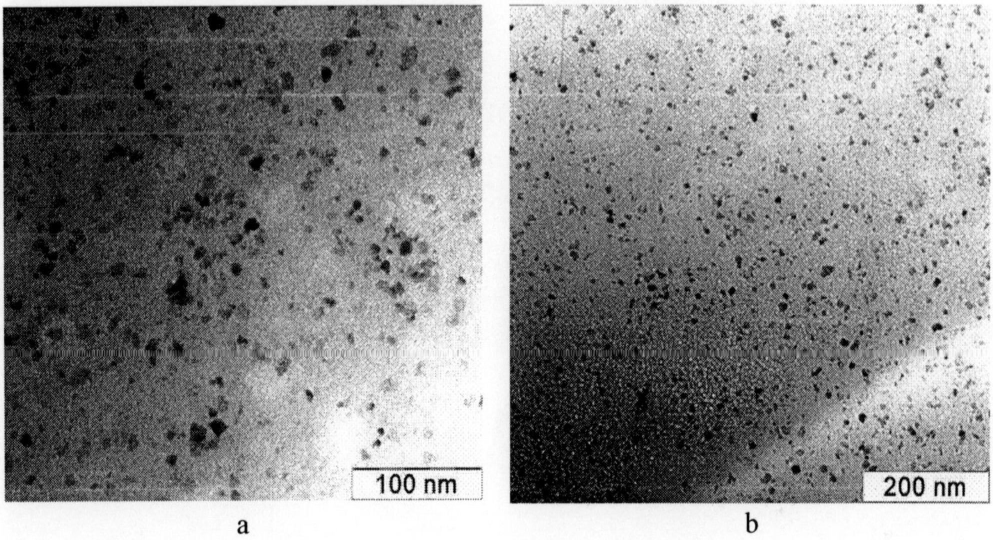

Figure 2. Electronic microphoto of ferrite nanoparticles in a magnetic fluid.

Figures 3 (*a* and *b*) show the Mossbauer spectra of sample S13 at 77 and 300 K, respectively. Th Mossbauer spectra were recorded in the transmission geometry, using an MCA (256 channels) and a Wissel constant acceleration transducer coupled to a 59 mCi ^{57}Co / Rh source. Each plastic sample holder (1.7 cm diameter) used to record the Mossbauer spectra contained 80 mg of uniformly distributed and pressed sample. The Mossbauer spectra in figures 3 (*a* and *b*) were least square fitted to a sextet and a doublet, giving parameters typical of cobalt ferrite [10]. At 77 K, the isomer shift (relative to natural iron) and the quadrupole splitting associated to the doublet are 0.47 and 0.71 mm / s, respectively. At 300 K, however, the isomer shift (relative to natural iron) and the quadrupole splitting associated to the doublet are 0.32 and 0.70 mm / s, respectively. At 77 and 300 K, the internal field associated to the sextet is 489.5 and 423.7 kOe, respectively. The collapsing of the internal field as the temperature increases from 77 to 300 K is typical of the superparamagnetic behavior of nanosized particles [11]. The Mossbauer doublet indicates cobalt-ferrite nanoparticles in the paramagnetic phase.

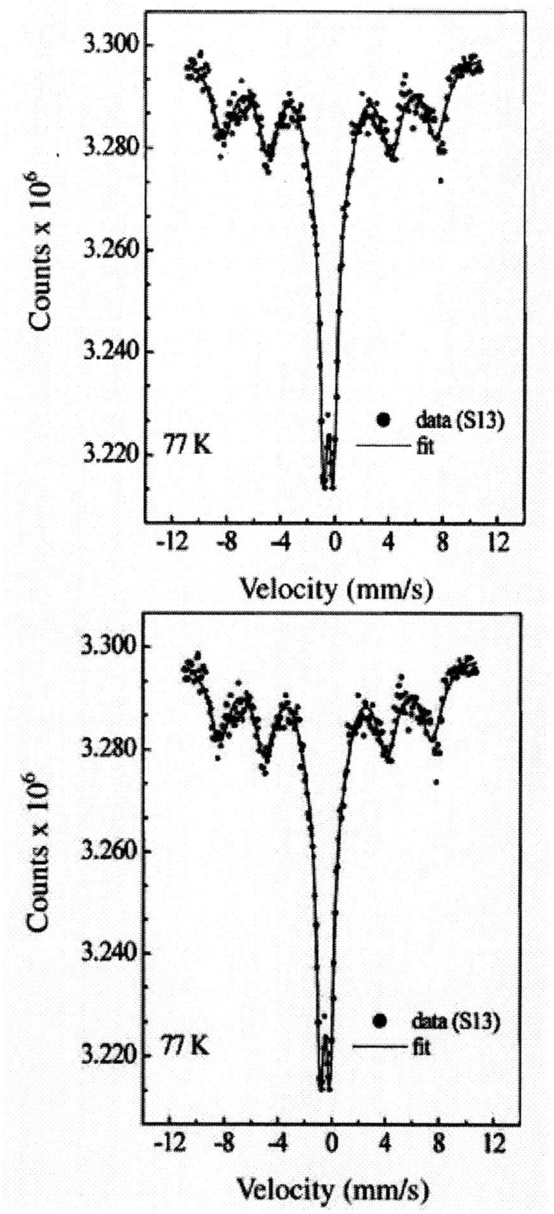

Figure 3. Mossbauer spectra of sample S13 at (a) 77K and (b) 300 K. Circles are the experimental data while the solid lines represent the best fit using one doublet and one sextet.

CONCLUSIONS

This work provides information concerning the synthesis of nanomagnetic particles, their passivation and peptization as stable ferrofluids. The most striking result is the possibility of controlling the nanoparticle size using different stirring speeds, with no apparent change of the standard deviation. The anomalous diffusion of the nanoparticles under turbulent stirring may provide the best picture for the size-control mechanism. Transmission electron microscopy (TEM) data show that increasing the stirring speed from 2700 to 8100 rpm reduces the average particle diameter from 15.1 to 11.4 nm, while keeping the standard deviation around 0.30. Mossbauer spectroscopy data identified the magnetic material as superparamagnetic cobalt ferrite, in agreement with the TEM data.

REFERENCES

[1] D. Gunther, N. Buske. DE patent № 4325386 (1993).
[2] D. Gunther, N. Buske. DE patent № 4372826 (1993).
[3] R. Massart. // *IEEE Trans. Magn.*, 17 (1981) 1247.
[4] B. M. Berkovsky, V. F. Medvedev, M. S. Krakov, Magnetic Fluids: Engineering Applications, Oxford University Press, New York, 1993.
[5] F. A. Tourinho, R. Franck, R. Massart. // *J. Mater. Sci.*, 25 (1990) 3249.
[6] S. Mitra, Applied Mossbauer Spectroscopy, Pergamon Press, Oxford, 1992.
[7] S. Chaves. // *Phys. Lett. A* 239 (1998) 13.
[8] D. del-Castillo-Negrete. // *Phys. Fluids,* 10 (1998) 576.
[9] B. Payet, D. Vincent, L. Delaunay, G. Noyel. // J. *Magn. Magn. Mater.*, 186 (1998) 168.
[10] H. R. Rechenberg, F. A. Tourinho. // Hyp. Interactions, 67 (1991) 627.
[11] H.- D. Pfannes, J. H. Dias Filho, J. L. Lopez. // *Hyp. Interactions,* 113 (1998) 507.

In: Preparation and Properties of Monomers, Polymers… ISBN 978-1-60021-557-5
Eds: A. Ballada and G. E. Zaikov, pp. 145-153

Chapter 16

IMMUNOMAGNETIC SEPARATION OF HUMAN HEMATOPOIETIC CELLS: PHYSICAL - CHEMICAL BASES AND MEDICAL - BIOLOGIC INVESTIGATION

F. S. Bayburtskiy*[*1], *N. A. Brusentsov*[2], *E. Yu. Filinova*[2], *E. A. Golenkina*[2], *P. K. Ivanov*[2], *D. Yu. Blokhin*[2], *V. I. Filippov*[2], *M. Yu. Larin*[2], *G. V. Stephanov*[3] *and O. L. Yershov*[3]

[1]N. M. Emanuel Institute of Biochemical Physics, Russian Academy of Sciences, 4, Kosygin str., Moscow 119991, Russian Federation

[2]N. N. Blokhin Russian Cancer Research Center RAMS, Kashyrskoye shosse, 24, Moscow, 115478 Russian Federation

[3]Research Institute of Chemistry and Technology of Hetero-Organic Compounds, Shosse Entuziastov, 38, Moscow, 111123 Russian Federation

ABSTRACT

Immunomagnetic separation of human hematopoietic cells is based on simultaneous use of strict selectivity of interaction monoclonal antibodies with membrane antigen's of cells - targets and a magnet management of carriers - magnetic microspheres. Cells selectively contacted particles become sensitive to a magnetic field and can be taken from an initial mix in a magnetic field. This not labor-consuming method allows processing the samples containing up to 10^{10} cells for short time. It is experimentally shown, that neither microparticles, nor procedure of separation do not render essential influence on viability, morphology and functional qualities of a biological material.

Keywords: immunomagnetic separation, human hematopoietic cells, monoclonal antibody.

---

[*] E-mail: Chembio@sky.chph.ras.ru, Bayburt@mail.ru

INTRODUCTION

Immunomagnetic separation is generally recognized as one of the most specific, reliable and fastest techniques to isolate cell subpopulations from complex cell mixtures [1 – 16].

The objective of our work was to develop the immunomagnetic reagent for human haematopoietic cell sorting. To this end monosized, superparamagnetic microspheres were generated and completely characterized. The conditions to conjugate monoclonal antibodies to the particles surface were optimized. The selection technique with the obtained immunomagnetic particles was further worked out.

MATERIALS AND METHODS

Preparation and Characterization of Magnetic Polystyrene Microspheres (MPM)

Magnetite (Fe_3O_4) prepared by the reaction of Fe (II) and Fe (III) with the ammonia aqueous solution was dispersed at organic solvent enriched with nonionic surfactant to obtain stable colloidal solution. Polystyrene magnetic particles with surface hydroxyl groups were prepared by the method of emulsifying polymerization of copolymers and magnetite ferrocolloid. Briefly, styrene (20 g), divinyl-benzene (2 g) and magnetite ferrocolloids (4,5 g) in styrene were emulsified in 400 ml of distilled water containing 0.25 % sodium dodecylsulfate (SDS) and agitated well for 10 min at room temperature. The mixture was ultrasonicated for 30 sec (frequency 22 kHz, power 40 W/cm^2). To perform polymerization 100 ml of 1% potassium peroxysulphate in distilled water was added and stirred at 60°C for 24 h followed by incubation at 70°C for 1 h. After reaching room temperature they were centrifuged at 500 g for 15 min. The sediment was resuspended in SDS and the magnetic particles concentrated in the magnetic field of a permanent magnet of magnetic intensity 2000 oersted for 30 min. SDS treatment was carried out twice. MPM were filtered with a Millipore 3 μm filter and resuspended in SDS for a final concentration of 10 mg/ml.

Particles were analyzed by transmitted-light microscopy. Size distribution was estimated by laser correlation spectroscopy. Magnetic properties (saturation and residual magnetization) were determined by magnetometry (Bruker, USA).

To increase the protein-binding capacity magnetic support was activated by *p*-toluenesulfonyl chloride. MPM were washed with water, water : dioxane (3:1, v / v), water : dioxane (1:3, v / v), dried dioxane. Wet weight treated particles (1g) were put into a round-bottomed flask containing 200 mg of tosyl chloride dissolved in 2 ml dioxane. Pyridine was added slowly under stirring. After 1 hour, the MPM were washed with dioxane and transferred back to water gradually. Activated particles were stored at 4°C until being used [1–4].

Binding of Antibodies to the Particles

Ammonium sulfate precipitation was used to obtain immunomagnetic conjugates murine IgG monoclonal antibodies (MoAbs) against human lymphocyte antigens CD3 (clone ICO-90), CD4 (clone ICO-86), CD8 (clone ICO-31) and CD20 (clone ICO-180) raised and purified from ascites fluids. Searching for optimal conjugation conditions we used varying buffer solutions (0.025M borate-buffered saline, pH 8.5; 0.05 M carbonate-buffered saline, pH 9.5; 0.05 M and 0.015 M phosphate-buffered saline, pH 7.4). Then we studied kinetics of immunoglobulin binding by altering the reaction time from 30 min to 24 h. Generally, to prepare conjugates 10 mg of corpuscular magnetic support (approximately $5 \cdot 10^{8}$ beads) were incubated with 200 µg MoAbs in a total volume of 1 ml buffer solution at 4°C under gentle rotation. Microspheres were then collected and washed twice using a magnetic field. Unoccupied proteinbinding sites were then blocked with 0.05 M PBS containing 2 % human serum albumin for 12 h at 4°C under gentle rotation. Antibody-coated magnetic particles were resuspended to a concentration of 10 mg/ml (approximately $5 \cdot 10^{8}$ particles per ml) with 0.01M PBS containing 2% human serum albumin (HSA) and 0.05% NaN_3 and stored at 4°C until being used.

Separation of Cells

Peripheral blood mononuclear cells (PBMC) from normal donors were isolated by density centrifugation on a Ficoll Paque gradient. Generally, immunomagnetic selection (depletion) of antigen-positive lymphocytes was carried out as follows: immunomagnetic particles were washed twice with 0.01M PBS (pH 7.4) using a magnetic field. $10 \cdot 10^{6}$ PBMC in 1 ml of 2 mM EDTA in 0.01 M PBS containing 0.5 % HSA were added to the collected particles to obtain an estimated 3:1 bead : target cell ratio. The resuspended mixture was incubated for 30 min at room temperature under gentle rotation. The bead / cell complexes were retained by the permanent magnet whereas unbound cells were removed by four washes with 0.01 M PBS. To evaluate depletion effectiveness flow cytometric analysis was carried out by direct immunofluorescence on samples from PBMS both before and after selection and on unbound cells (negative fraction). Data acquisition and analysis were assessed on a FACScan using FACScan research software (Becton Dickinson and Co.) [5 – 8].

To determine an optimal quantity of immobilized MoAb magnetic beads conjugated with anti-CD ^{3} and anti-CD $^{20-}$ antibodies that were known to have bound 4, 7, 10 and 12 µg of antibody per 1 mg of support were used, the results obtained were analyzed for different processing variables.

Results and Discussion

MPM Preparation and Analysis

Our preparation method for the magnetic polystyrene microspheres formed spherical particles (Figure 1). Laser correlative spectroscopy showed that the particles had a mean diameter of 1.0 μm (mkm) with size homogeneity of more than 90 %.

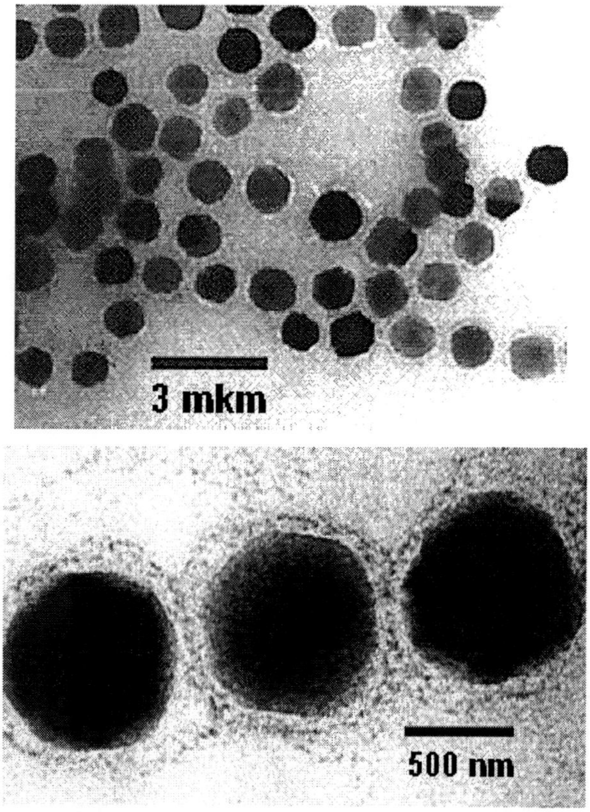

Figure 1. Transmitted-light microscopy of synthesized magnetic polystyrene particles.

Estimated basic magnetic properties are given in Table 1. Due to the insignificant residual magnetization, MPM can be repeatedly collected in a magnetic field and there after easily redispersed. The saturation magnetization of 25 Gs · cm ^{3} / g enabled the rapid concentration of the MPM using a permanent magnetic field of 2000 Oe.

Table 1. Measured magnetic characteristics of MPM synthesized

Remanent magnetization, Gs · cm ^{3} / g	Saturation magnetization, Gs · cm ^{3} / g
1.2	25.0

Coupling of Antibodies to the Particles

MPM did not bind MoAbs when incubated in 0.05 M carbonate-buffered saline at pH 9.5. The maximum protein-binding capacity was less then 2 μg MoAbs per mg particles when 0.025 M borate-buffered saline at pH 8.5 or 0.05 M phosphate buffered saline at pH 7.4 were used. We elicited 0.015M PBS at pH 7.4 to be best for immunoglobulin coupling to the particles. In this way, 10 – 15 μg MoAbs per mg particles could be immobilized after incubation of the MoAbs / MPM mixture for 24 h. When studying the kinetics of immunoglobulin coupling we discovered that the binding rate was not the same for different MoAbs (Figure 2). The optimal antibody-binding capacity as determined by subsequent immunomagnetic selection was found to be 7 – 10 μg MoAb per mg particles. The required incubation time was averaged to be 10 hours for conjugating with clones ICO-90, ICO-86, ICO-31, and not more than 6 hours for clone ICO-180.

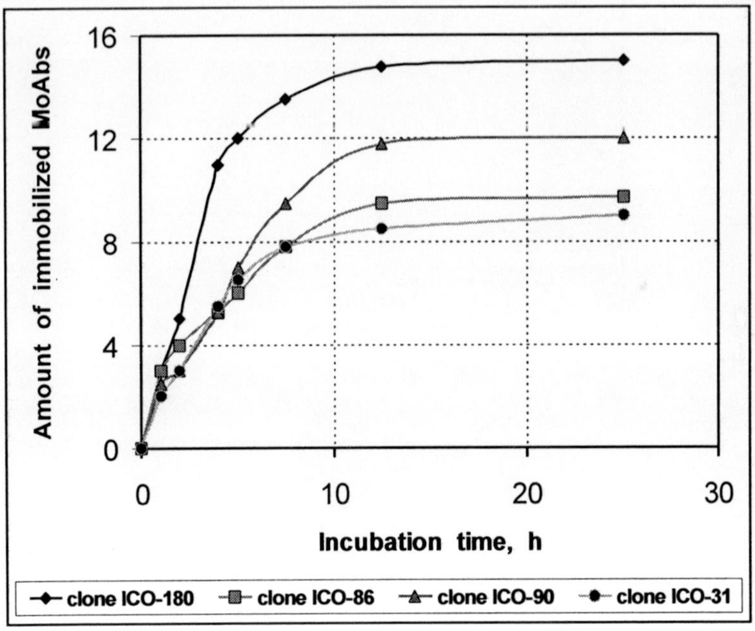

Figure 2. Rates of IgG immobilization and MPM surface. Amounts of immobilized immuno-globulines (μg per 1 mg support) (Y) were measured as a function of interaction time (h) (X) for different MoABs clones.

Immunomagnetic Cell Selection (Depletion)

Our primary concern was to determine an optimal amount of MoAbs that should be immobilized on the particles surface. For this purpose, a series of CD $^{3+}$ – lymphocytes depletions were carried out. As noted before, immunomagnetic anti-CD^3 conjugates with 4 to 12 μg IgG per mg support were used. There proved depletion grade to be estimably influenced by the level of coupled MoAbs, the other conditions (i. e. source PBMS, beads per cell ratio, incubation time and temperature) being equal. The antibody binding capacity of 4 μg / mg was apparently deficient because in this case immunomagnetic selection failed to eliminate CD^{3+} targetcells. The maximum antibody binding capacity of 12 μg per mg support for clone ICO-90 was also not optimal, because excess IgG split out easily and blocked CD^{3} receptors on target cells interfering with bead / cell complexes formation. The latter was detected by staining of unbound cells after selection with FITC labeling sheep anti-mouse F(ab)2 [9 – 12]. The most efficient was the depletion with MPM covered with MoAbs in concentration 7 to 10 μg. Negative selection in those cases resulted in depletion of 97 % CD^{3+} – lymphocytes MoAbs (Figure 3).

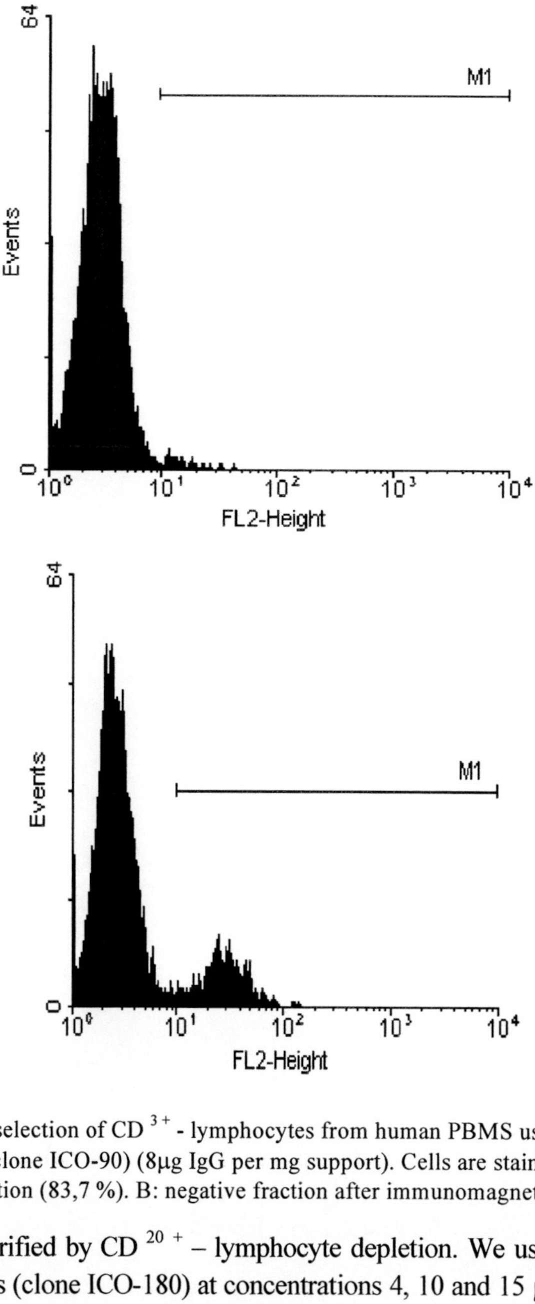

A: 83,7 %

B: 1,9 %

Figure 3. Negative selection of CD^{3+} - lymphocytes from human PBMS using MPM conjugated with MoAbs anti-CD^{3} (clone ICO-90) (8μg IgG per mg support). Cells are stained with CD^{3} FITC. A: PBMC before selection (83,7 %). B: negative fraction after immunomagnetic depletion (1,9 %).

Data were verified by CD^{20+} – lymphocyte depletion. We used MPM conjugated with anti-CD^{20} MoAbs (clone ICO-180) at concentrations 4, 10 and 15 μg per mg of support. IgG

ICO-180 concentrations of 10 µg / mg were optimal, providing a depletion rate of 96 % (Figure 4). For further experiments immunomagnetic particles covered with MoAbs anti-CD^{4} (clone ICO-86) and anti-CD^{8} (clone ICO-31) were used at a concentration of 8 µg/mg. Immunomagnetic separations with these conjugates eliminated 95 – 97 % of antigen positive cells from an initial PBMS suspension (data not shown).

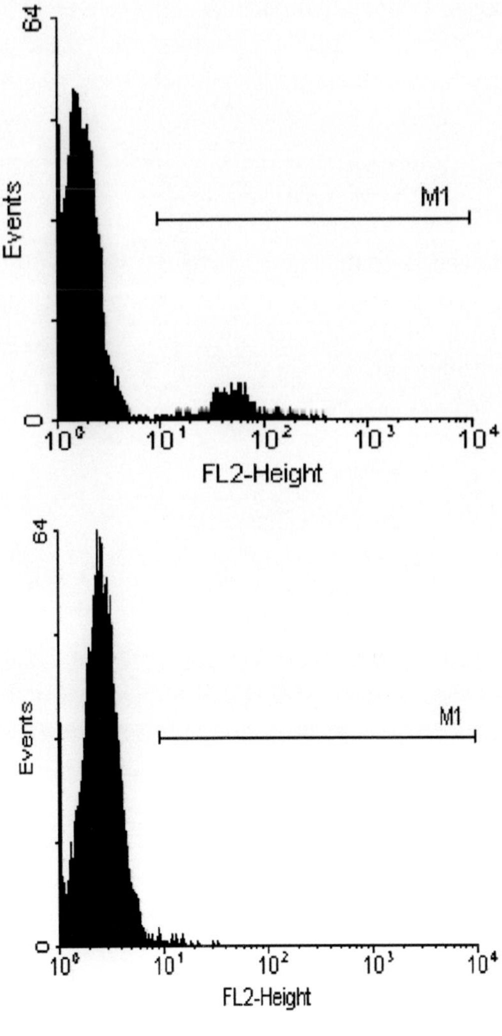

A: 6,6 %

B: 0,2 %

Figure 4. Negative selection of CD^{20+} - lymphocytes from human PBMS using MPM conjugated with MoAbs anti-D20 (clone ICO-180) (10 µg IgG per mg support). Cells are stained with CD^{20} PE. A: PBMS before selection (6,6 %). B: negative fraction after immunomagnetic depletion (0,2 %).

Polystyrene magnetic microspheres were prepared from copolymers and magnetic iron oxide employing the method of emulsifying polymerization. The resulting particles were 1µm in diameter, the homogeneity level was better than 90 %. Microspheres possessed insignificant magnetic remanence and showed superparamagnetic behavior. To provide active surface groups MPM were activated by *p*-toluenesulfonyl chloride [13 – 16].

Optimal immunoglobulin coupling for the particles was performed in 0.015 M PBS at pH 7.4 at 4°C with gentle rotation. The interaction time should be determined for each MoAb clone because the immobilization rate is not constant. To achieve the optimal amount of surface-bound IgG, which was estimated to be 7 – 10 μg MoAbs per mg support, conjugation times had to be 6 h for clone ICO-180, and 10 h for clones ICO-90, ICO-86 and ICO-31. Both deficient and excessive immunoglobulin coupling was proven to be inadvisable. Immunospecific conjugates against human lymphocyte antigens CD ^{3}, CD ^{4}, CD ^{8} and CD20 were consequently generated. Immunomagnetic separation with these conjugates is able to eliminate about 97% of antigen positive cells from an initial PBMC suspension.

Conclusions

We believe that the developed immunomagnetic reagents will be useful in laboratory and clinical practice. They might be especially useful for the removal of tumor cells from bone marrow before autologous transplantation (in patients with B- or T-cell lymphomas). They also could be applied for T-lymphocyte purging of bone marrow before allogeneic transplantation to avoid development of graft versus host disease.

References

[1] K. Nilsson, K. Mosbach. // *Eur. J. Biochem.* 112 (1980) 397 – 402.

[2] Rembaum, S. Margel // *Br. Polym. J.* 10 (1979) 275 – 280.

[3] E.A. Golenkina, P.K. Ivanov, E.R. Polosukhina, O.N. Donskaya // *High Medical Technologies, Materials of the V Russian Oncological Congress* (2000) 243 – 244.

[4] B. Brockstein, A. Ross, K. Hollingsworth // *Proc. Am. Soc. Clin. Oncol.,* 14 (1995) 327.

[5] R. Chakraverty, S. Robinson, K. Peggs. // *Bone Marrow Transplant.,* 28 (9) (2001) 827–834.

[6] L. Garderet, V. Snell, D. Przepiorka. // *Transplantation*, 67 (1) (1999) 124–130.

[7] R. Handgretinger, J. Greil, U. Schirmann, P. Lang, O. Gonsalez-Ramella, I. Schmidt, D. Niethammer, T. Klingebiel. // *J. Hematother.,* 6 (3) (1997) 235-242.

[8] C. Herrera, J. M. Garcia-Castellano. // *Bone Marrow Transplant.,* 23 (5) (1999) 443 – 450.

[9] S. Karlsson, P. H. Correll, L. Xu. // *Bone Marrow Transplant.,* 1 (1993) 124.

[10] J. A. O'Shaughnessy, K. H. Cowan, A. W. Nienhuis, K. T. McDonagh. // *Hum. Gene Ther.,* 5 (1994) 891.

[11] M. Ogura, Y. Kagami, R.. Suzuki. // *Cancer Chemotherapy Pharmacology,* 40 (1997) 51–57.

[12] H. Sao, K. Kitaori, M. Kasai // Int . J. Hematology, 69 (1) (1999) 27 – 35.

[13] F. Servida, D. Soligo, L. Caneva. // *Stem. Cells,* 14 (4) (1996) 430 – 438.

[14] R. J. Soiffer, E. Weller, E. P. Aliea. // J. Clin. Oncol., 19 (4) (2001) 1152 – 1159.

[15] M. Ward, C. Richardson , P. Pioli, L. Smith, S. Podda, S. Goff. // *Blood,* 84 (1994) 1408.

[16] L. Xu, S. K. Stahl, H. P. Dave, R. Schiffmann, P. H. Correll. // Exp. Hematology, 22 (1994) 223.

In: Preparation and Properties of Monomers, Polymers... ISBN 978-1-60021-557-5
Eds: A. Ballada and G. E. Zaikov, pp. 155-163

Chapter 17

EMULSION POLYMERIZATION OF (METH)ACRYLATES: CHARACTERISTICS OF KINETICS AND MECHANISM

N. V. Kozhevnikov, M. D. Goldfein, A. V. Trubnikov and N. I. Kozhevnikova
Saratov State University, 83, Astrakhanskaya st.,
Saratov 410012, Russia

ABSTRACT

Kinetics and mechanism of emulsion polymerization of methylacrylate and methylmethacrylate are investigated. It is shown that kinetics behaviour of the reaction is determined by: different nucleation mechanisms, initiation of gel-effect, bimolecular chain termination in water phase, floculation of polymer-monomer particles during all polymerization stages, partial emulsifier solubility in the monomer. These effects lead to the dependence of particles number and reaction rate in particles on conversion and to the influence of polymerization conditions on kinetics orders with respect to initiator and emulsifier concentrations.

INTRODUCTION

Emulsion polymerization remains one of the most problematic sections of radical polymerization as a whole. Development of its general theory meets principal difficulties. The reasons are multi-phase structure of emulsion system, variety of parameters determining kinetics and mechanism of the process. These parameters depend not only upon reagents' reactivity, but also on their phase distribution, reaction topochemistry, mechanism of particles' nucleation and stabilization.

The classic emulsion polymerization theory [1] was developed for monomers with low water solubility (styrene). It postulates that polymer formation takes place in the polymer-monomer particles (PMP), which are formed from emulsifier micelles with solubilized monomer inside. Transformation of micelles into PMP occurs when they are entered by radicals generated in the aqueous phase due to the decomposition of water-soluble initiator (micellar nucleation). In the case of monomers with higher water solubility (acrylates), the particles can be formed according to homogeneous nucleation mechanism [2], when oligomeric radical, growing in water, loses its solubility and precipitates, forming the PMP nucleus. Polymerization rate is determined by the rate of the reaction in separate particles and their number. According to the classic concepts, at each moment PMP can contain not more than 1 radical, because when the second one is entering PMP, chain termination occurs immediately. Thus the average number of radicals in the particles $\overline{n}$ is equal to 0.5.

EXPERIMENTAL

The present paper studies the emulsion polymerization of methylacrylate (MA) and methylmethacrylate (MMA), initiated by ammonia persulphate (PSA). Monomers were purified by distillation at lowered pressure in the presence of inert gas flow, PSA – by recrystallization. Emulsifiers – sodium lauryl sulfate (LS), sulfated oxyethylated alkylphenoles – S-10 and its analogue Neonol $AP_{9-12}S$. Polymerization was carried out in helium atmosphere. Monomer and water solution of initiator and emulsifier were freed of oxygen beforehand by thrice-repeated procedures of freezing, high-vacuum air exhausting and defrosting in vacuum. Reaction kinetics was followed by dilatometry, latex properties (particles size and number) were studied by turbidity spectrum method [3] (spectrophotometer SPh-26). The dependence of particles' spectral properties on conversion and refraction index dispersion were taken into account [4].

RESULTS AND DISCUSSION

Emulsion polymerization goes through 3 main stages – PMP formation (when free emulsifier is available), stationary process (when monomer droplets in water phase are present, and monomer equilibrium concentration is established in PMP), and reaction completion (when monomer in PMP is depleted). For many monomer–polymer systems the parameters of equilibrium swelling of latex particles are well- known [5]. Such parameters indicate that in the case of polar monomers droplets' disappearance and completion of constant rate sections on kinetic curves must occur at relatively low conversions. For example, in the case of MA polymerization – at 16 % conversion, MMA – 34 % [5].

Our research showed that constant rate stage continues to the higher conversions q, depending on monomer structure and reaction conditions. In some cases polymerization rate does not slow down at high conversions, but, on the contrary, increases. Such effects, occurring notwithstanding the fact that monomer concentration in PMP is decreased, indicate that average radicals' amount is growing up, which is possible when large enough particles are formed and when high viscosity takes place in PMP [6].

Simultaneous growth of several radicals in the particle leads to polymerization acceleration and initiation of the so-called "gel-effect". The probability of gel-effect, according to [7], can be judged by the value of parameter α, which is defined as:

$\alpha = \vartheta V / k_o N$, where ϑ – the total rate of radical entrance to the particles, N and V – their number and average volume, k_o – rate constant of bimolecular chain termination. Polymerization rate growth, caused by increasing of k_o, starts to be noticeable when $\alpha > 10^{-4}$, at $\alpha > 10^{-2}$ gel effect is pronounced. If, as a first approximation, to assume that ϑ is equal to the rate of radicals formation in water phase (initiation rate), and use the final dispersion data, than α parameter value can be estimated. The results are shown in the figure 1. It indicates the existence of gel-effect, which is especially strongly pronounced when Neonol $AP_{9\text{-}12}S$ and S-10 are used as emulsifiers. In the presence of these emulsifiers larger latex particles are formed, which promotes the origination of gel-effect. It weakens while emulsifier concentration is increased, but becomes stronger with the growth of initiation rate.

In the case of MA polymerization gel-effect starts already at initial reaction stages. MMA is characterized by lower propagation rate constant, and because of that conditions for gel-effect arise only at high enough conversion, when monomer concentration decreases and viscosity in PMP grows up. This leads to more pronounced influence on the shape of kinetic curves.

Thus, for the emulsion polymerization of studied monomers classical conceptions of instantaneous chain termination in PMP in the moment of second radical entrance are not valid.

The growth of emulsion polymerization kinetic order with respect to emulsifier (LS) with the increase of initiator concentration is reported in [8]. This is connected to the interaction of oligomeric radicals and chain termination in the water phase. In the present work, studying reaction rates, we also found similar relationships (figure 2).

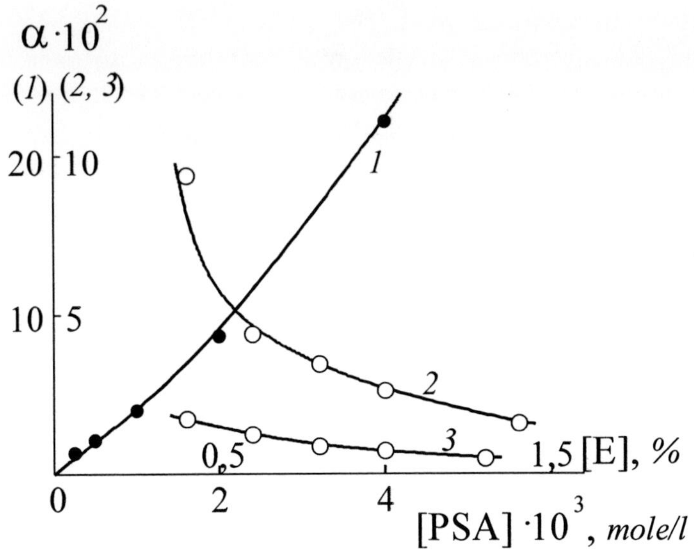

Figure 1. Parameter α value against initiator concentration (*1*), emulsifier Neonol $AP_{9\text{-}12}S$ (*2*) and LS (*3*) concentrations, emulsion polymerization of MA. (Parameter α is the measure of gel-effect probability). [Neonol] = 1 %, 60 °C (*1*); [PSA] = 5 10^{-3} mole/l, 50 °C (*2, 3*).

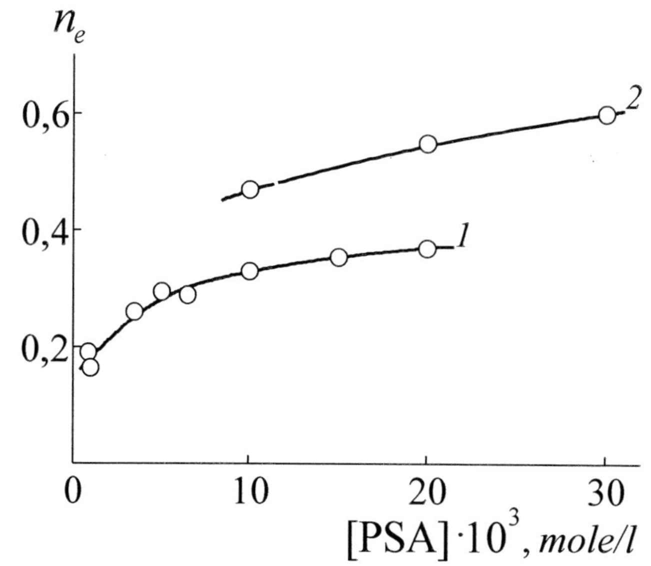

Figure 2. Kinetic order of MA emulsion polymerization with respect to emulsifier LS (*1*) and Neonol $AP_{9-12}S$ (*2*) versus initiator concentration. [MA] = 20 %; 50 °C.

Moreover, kinetic order with respect to emulsifier n_e is greater in the case of S-10 or Neonol, than with LS. This is the result of partial solubility of oxyethylated alkylphenols in the monomer. We found (using electronic spectroscopy) that with the growth of these emulsifiers' concentration the part of them staying in water also increases. This part is promoting PMP formation, which increases the influence of Neonol and S-10 concentration on polymerization rate comparing with LS.

Established dependences of kinetic order with respect to emulsifier on reaction conditions, emulsifier properties indicate that n_e, contrary to the prevailing view, cannot characterize the emulsion polymerization of certain monomer.

The classic emulsion polymerization theory quantitatively describes the influence of emulsifier and initiator on the number of formed latex particles. Including the reaction rate into the same description pattern presumes that it is proportional to the number of particles. However, the reaction rate is determined not only by number of particles, but also by the rate of monomer conversion inside them, which can change due to the conditions and depth of polymerization (for example, due to different degree of gel-effect).

Our studies of emulsifier concentration effect on latex particles number N_{100} in the final dispersion (q =100 %) showed, that kinetic order with respect to emulsifier, obtained using this data, is much larger than calculated using rate values (especially in the case of Neonol and S-10).

It's important to keep in mind that emulsion polymerization rate and N_{100} value characterize the process and forming dispersion at different conversions and comparison between them is justified only when one of the basic postulate of classical theory is working – number of PMP during the reaction after completion of its first stage is constant. However, our research showed that number of particles in the emulsion depends on polymerization depth. The most common effect – decreasing of particles number during the third stage of the reaction (figures 3, 4).

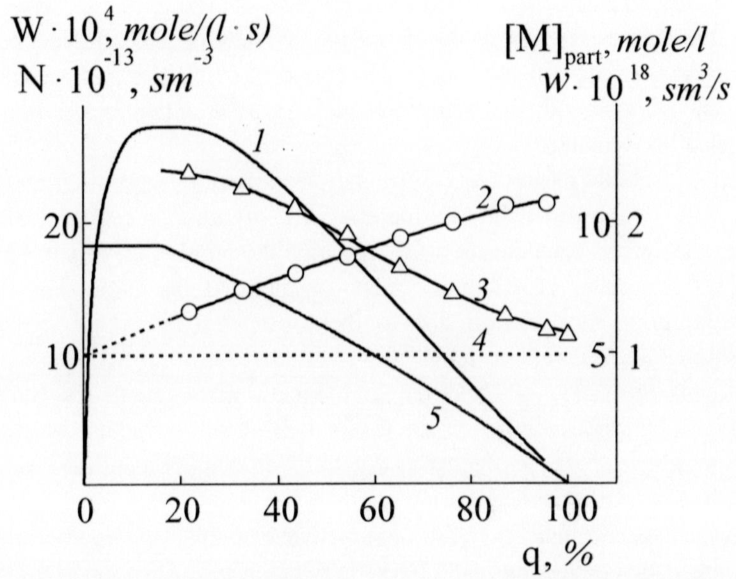

Figure 3. Polymerization rate (*1*), number of particles in emulsion (*3*), monomer concentration and rate in the particle (*5*, *2*) against monomer-polymer conversion depth, emulsion polymerization of MA. *4* – Specific rate in the particle at $\overline{n}$ = 0,5. [PSA] = 0,25·10^{-3} mole/l; [LS] = 1 %; 60 °C.

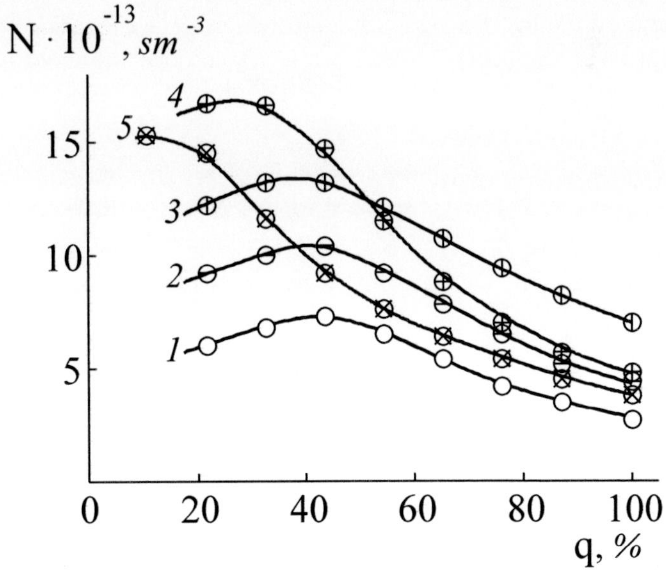

Figure 4. Number of particles in the emulsion versus monomer–polymer conversion depth, polymerization of MA. [PSA]·10^3 = 0,1 (*1*), 0,25 (*2*), 1 (*3*), 4 (*4*), 8 mole/l (*5*); [Neonol] = 1 %; 60 °C.

This indicates the changing of interphase surface condition and decreasing of its stabilization degree with the conversion growth. Such effects can be caused by various reasons, for example by changing the ratio of monomer/polymer concentrations in the particle. In addition, all the time during polymerization new and new charged oligomer radicals are entering PMP from water phase and sometimes start the reaction in the particle, sometimes terminate it. Such radicals are increasing the charge on the particles' surface and strengthening the electrostatic repulsion of surface active substances molecules (SAS), that hinders their absorption and can cause even desorption of emulsifier. As a result, floccullation of particles occurs, not only at the initial conversions, as assumed in [5], but also at more advanced stages of the reaction.

When partially soluble in monomer S-10 and Neonol $AP_{9-12}S$ are used as stabilizers, the conversion growth is followed at first by increase of PMP number, and only after that by its decrease (figure 4). In this case during polymerization the emulsifier, dissolved in monomer, is released. This emulsifier is stabilizing new particles. At the same time the emulsifier replenishment promotes flocculation due to the growth of interphase surface, and also because new small particles are less stable [9].

Flocculation is the cause of abnormal high kinetic orders with respect to emulsifier (>0,6), found from particles number data in final dispersion, because at high emulsifier concentrations number of particles decreases due to the flocculation at third reaction stage not so heavily as when emulsifier concentration is low.

We determined specific reaction rate in the particles, comparing reaction rate and number of particles at the same conversion level. Specific reaction rate was found to change with the conversion. This is related to a considerable degree to the decreasing of monomer concentration in the particle $[M]_{part}$. At the initial polymerization stages, when monomer droplets are still present in the reaction system, monomer concentration in the particle is determined by equilibrium degree of dissolving in polymer (equilibrium swelling degree) and remains constant. At higher conversions it decreases. If we consider all monomer and polymer to sit in PMP, the dependence of $[M]_{part}$ (mole/l) on polymerization depth at third reaction stage will be given by:

$$[M]_{part} = \frac{1000 \cdot \rho_M}{m_o} \cdot \frac{1-q}{1-q \cdot (1-\rho_M/\rho_p)}$$

(ρ_M, ρ_p – densities of monomer and polymer, m_o – monomer molecular mass).

Specific rate in the particle $w = W/(N\,[M]_{part})$ grows with monomer–polymer conversion degree (figures 3, 5). This means that radicals' number in PMP is increasing. Moreover, the extrapolation of specific rate to the initial conversion gives the value corresponding to the specific rate at $\bar{n} = 0{,}5$ (dotted line *4* in figure 3 is calculated assuming k_p for MA is equal to 1190 l/mole · sec [10]). Reaction rate in the particles grows with the conversion depth depending on initiator concentration – higher concentrations lead to higher growth (figure 5), indicating the increase of gel-effect. Consequently, our data confirm the previous conclusion that in the case of MA emulsion polymerization gel-effect emerges already at low conversions and than grows gradually as particle size in increasing at second polymerization stage, and as viscosity is growing at third stage due to increasing of polymer concentration. When MMA is polymerized, gel-effect starts to be noticeable only at high conversions and is much more manifested than with MA (figure 5).

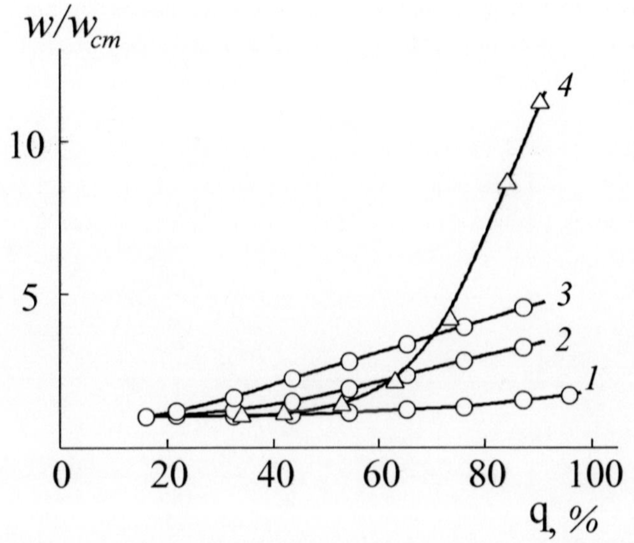

Figure 5. Behavior of specific reaction rate in the particle at the third stage of MA (*1–3*) and MMA (*4*) emulsion polymerization. [PSA]·10^3 = 1 (*1*), 1,5 (*4*), 4 (*2*), 8 mole/l (*3*); [Neonol] = 1 (*1–3*) и 2 % (*4*); 60 °C.

Initiation rate also influences emulsion polymerization of (meth)acrylates in an unusual way. The parameters of this process and polymer dispersion differently depend on initiator concentration. The maximal rate value W_{max}, as a rule, increased with the growth of PSA, but differently in the various ranges of PSA concentration change. As a result, the graph of initiation rate versus initiator concentration in logarithmic scale (used to calculate the order with respect to initiator n_i) has the form of broken line with the point of inflexion (figure 6). Value of n_i at the relatively low initiator concentrations turned out to be higher than expected theoretical value of 0,4 (n_i= 0,6 – 0,7), but in the relatively high concentration region – lower (n_i= 0,25 – 0,35). Thus, kinetic order of the emulsion polymerization with respect to initiator depends not only on monomer properties, nature and concentration of emulsifier, but also on the region of initiator concentrations where it is measured.

Number of particles in the polymer dispersion N_{100} is growing initially, but later starts to decrease with the growth of PSA concentration in the presence of Neonol or S-10. The stationary rate curve also passes maximum. The observed extremal curves are related to the flocculation processes, taking place in the varius polymerization stages. For instance, dependence of N_{100} on intiator concentration is determined by flocculation at all stages of the process, while stationary rate is influenced by flocculation only at initial stages. Behavior of maximum rate versus [PSA] is influenced both by flocculation growth and by initiator concentration increase, which promotes acceleration of the reaction in the particles. Therefore, at relatively low initiator concentrations, when flocculation is still weak, the gel-effect intensification causes more drastic dependence of reaction rate on initiator concentration than is expected based on classic concepts (n_i> 0,4). At the high initiator content, depending on flocculation – gel-effect relative efficiency correlation, reaction rate continue to grow (though with low n_i value), or even starts to decrease.

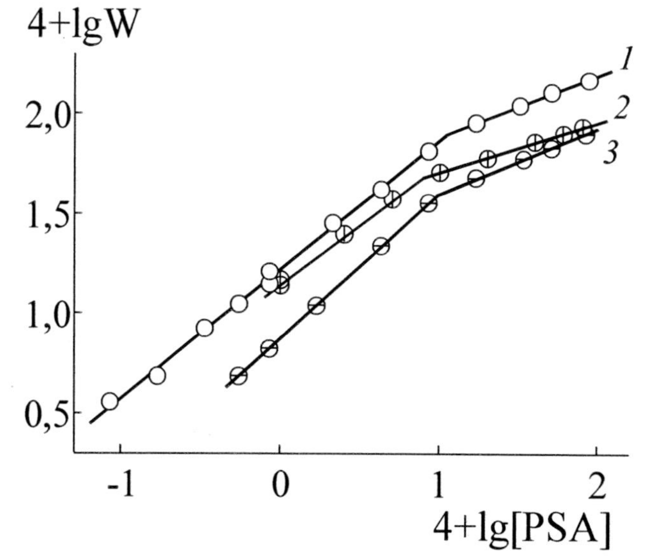

Figure 6. Emulsion polymerization of MA: rate against initiator concentration. Emulsifier – LS (0,4 %) (*1*), Neonol (1 %) (*2*), S-10 (1 %) (*3*). 60 °C.

Number of particles in emulsion at the conversion corresponding to maximum polymerization rate also depends extremally on initiator concentration. This fact, together with the stationary rate passing maximum, indicates that flocculation takes place at the initial reaction stages. At high [PSA] flocculation completely compensates the growth of particles' number due to the initiation rate increase; at low [PSA] this is not the case. Thus, flocculation at initial polymerization stages increases with the initiation rate growth and reduces the n_i value. Flocculation at the third stage, as a rule, almost does not influence W_{max} and, consequently, n_i, because the rate reaches maximum at lower conversions. However, it reduces the number of particles in final dispersion. The reducing effect is more striking with emulsifiers S-10 or Neonol, than with LS, because the first two emulsifiers cause stronger flocculation at the third reaction stage, due to emulsifier replenishment from monomer droplets or PMP. In the case of LS N_{100} value is increasing in all studied initiator concentration range and does not pass maximum.

References

[1] Smith W.V., Ewart R.H. // *J.Chem. Phys.*, 1948. Vol. 16. № 6. P. 592.

[2] Hansen F.K., Ugelstad J. // *J. Polymer Sci.: Polymer Chem. Ed.*, 1978. Vol. 16. № 8. P. 1953.

[3] Klenin V.I., Schegolev S.U., Lavrushin V.I. *Characteristic functions of light scattering in disperse systems,* Saratov, 1977.

[4] Kozhevnikov N.V., Kozhevnikova N.I., Goldfein M.D. // *Journal of Applied Spectroscopy*, 2005. Vol. 72. № 3. P. 313.

[5] Eliseeva V.I. *Polymer dispersions*, Moscow, Khimiya, 1980.

[6] Ivanchev S.S. *Radical polymerization*, Leningrad, Khimiya, 1985.

[7] Friis N., Hamielec A.E. // *Amer. Chem. Soc. Polymer Preprints,* 1975. Vol. 16. № 1. P. 192.

[8] Kozhevnikov N.V., Goldfein M.D., Zubin B.A., Trubnikov A.V. // *Vysokomolek. Soedin,* A. 1991. Vol. 33. № 6. P. 1272.

[9] Dunn A.S., Chong L.C-H. // *Brit. Polym. J.* 1970. Vol. 2. № 1-2. P. 49.

[10] Bagdasaryan H.S. *Theory of Radical polymerization*, Moscow, Nauka, 1966.

In: Preparation and Properties of Monomers, Polymers… ISBN 978-1-60021-557-5
Eds: A. Ballada and G. E. Zaikov, pp. 165-173

Chapter 18

BEHAVIOR OF COMPOSITE MATERIALS UNDER MICROORGANISMS OF SOIL

O. A. Legonkova* and O. V. Selitzkaya**

*Moscow State University of Applied Biotechnology,
33, Talalikhina str., Moscow 109316, Russia
**Russian State Agrarian University named after K.A. Timirjasev,
48, Timirjazevskaya str., Moscow 127550, Russia

ABSTRACT

The article is devoted to the investigation of the dynamics of fungi and bacteria growth in different types of soil in dependence on the presence of composite materials based on polymers.

Keywords: composite materials, soil microorganisms, biodegradation, polymers.

INTRODUCTION

Today polymers and composite materials on polymer base occupy a leading place in our every day life. To preserve of oil resources and to reduce of wastes a lot of biodegradable materials with the improved quality characteristics are being created [1].

Biodegradable polymer materials can be divided into three groups: plastics based on natural biopolymers; chemically and microbiologically synthesized polymers and their mixes; composite materials [2], which contain biodegradable components, such as: cellulose and its esters, starch and etc. that lead to bio-damage of the whole composition. These materials are of special interest for practical utilization [3, 4].

* E-mail: OALegonkovaPB@mail.ru

Stability to microorganisms action of polymer materials depends on the used softeners, fillers, stabilizers, different technological additives, depends on their ability to be a source of carbon, nitrogen and other biogenic elements.

Successful decision of the problem connected with bio-decomposition of composite materials is closely related with the evolution of submission of the nature of the real processes of microbiological damages and presence of objective quantitative information about the regularity of their arising and proceeding.

The aim of the present work was to investigate the behavior of composite materials on the base of series of large-capacity plastics under composting conditions for creation of ecologically safe materials with different terms of utilization.

EXPERIMENTAL PART

The following materials were taken as polymer basis: acrylate-styrene carboxylated latex – Lentex A4; co-polymer of ethylene and vinyl acetate (CEVA); co-polyamide (Co-PA), received by polycondensation of adipinic and sebacic acids and hexamethylendiamine, trade mark – H-005 (Tg = 115-120^{0}C, melting index – 15 g\10min); thermoplastic polyurethane; polyvinyl alcohol (PVC). The criterion, governing the chose of the said above polymer materials, was the presence of functional groups in all these polymers. Another determining factor was the low temperature of their processing (130-150^{0}C). This factor is rather important from the point of preservation of properties of fillers during getting samples [5].

Organic and inorganic fillers were taken. Waste of grain thrashing were chosen as an organic filler, it has particles of 63-240 mkm in dimension, piled density – 350 kg/m^3 , moisture – 4%. Water soluble mineral complex fertilizer of type "Rastvorin-A" was used as an inorganic filler which consists of the necessary microelements needed for growth and evolution of plants as well as the main trace elements, such as: Mn, Cu, Zn, B, Mo, in the form of their salts, table 1.

Table 1. Qualitative characteristics of "Rastvorin A"

Nitrogen, in general, %	10,0
Including N-NH_2	-
Including N-NH_4	5,0
Including N-NO_3	5,0
P_2O_5, %	5,0
K_2O, %	20,0
MgO, %	5,0
pH	3,0-4,5
Not soluble residue	less than 0,1

In order to compare the behavior of the prepared samples under composting conditions, composite materials were placed into soils of two kinds: sod-podzol soil (#1) and hothouse peat ground (#2). The samples were kept at 25^{0}C during 5 months. The results were received with the help of the following methods:

- methods of analytical selection. Microbiological investigations were directed on quantitative determination and definition of domestic forms of microorganisms [6]. Layers of soil were investigated which directly surrounded the composted samples;
- chromatography-mass-spectrometry (Hewlett Packard HP – 6890 chromatograph, USA);
- scanning electron microscopy (microscope JOEL - JSM-5300LV, Japan).

RESULTS AND DISCUSSIONS

It's well known fact that there is no absolutely stable to action of microorganisms polymers. Plastics are damaged by the following germs: fungi - *Aspergillus niger, Asp. versicolor, Asp. flavus, Asp. amstelodamii, Asp. ruber, Pen. purpurogenum, Pen. brevi-compactum, Pen. commune, Penicillium palitans, Cladosporium, Fusarium, Paccllomyces Penicillium sp., Aspergillus wamori, Aspergillus oryzae, Trichoderma sp., Chaetomim globosum, Trichoderma lignorum, Cephalosporum aeremonrum, Rhizopus nigricans, Alternaria, Candida, Lipolitica, Sporatrichum sp.,* bacteriums – *Ptreptococcus rubrireticuli, Pseudomonas aeruginosa, Streptococcus rubrireticull* [7-15].

It was revealed by us that the main part of the grown fungi on organic filler consists of *Penicillium*, also *Aspergillius (Asp.glaucus, Asp. restrictus, Asp. halophylicus, Asp. candidis)* and *Thrihoderma* were present.

Fungi of *Penicillium u Aspergillius* genus are the most important types of soil microorganisms that influence mostly on preservation and quality of seeds and grains' waste. For that reason they are called fungi of storage or moulds of storage (while growing they oppress a lot of other fungi, for example *Cladopsorium, Trihcothecium)*. Thus, composite materials, containing organic filler, have been already contaminated with fungi spores, arousing bio-damage of polymer composite materials while being composted.

In the tables 2-4 the results on changing of microorganisms quantity in two soils (#1 and #2) while polymer composite samples composting during different testing times (from 2 to 5 months) are presented.

As it's shown during the first two months the fungi quantity on Chapek medium for fungi is the sample within the bounds of experiment and doesn't depends on polymer basis. Under aerobic conditions fungi occupy the significant role in waste of grains decomposition [16], all that is confirmed during this experiment.

The quantity of bacterium and fungi in soil #1, with smaller amount of humus, is higher then in the soil #2. Besides microorganisms (*Penicillium*, *Aspergillius, Thrihoderma*), being peculiarly present in organic filler, fungi of *Mycor, Rhizopus, Fusarium* genus were revealed which are characterized not only with low growth but the ability to decompose aroma compounds.

Attention should be paid to the fact that a lot of actinomycetes have appeared after 5 months of composting in all soils. We haven't mentioned them during the earlier periods of composting. It's known that formation of humus can take place while suppressing the microbiological processes [17]. Thus, in our case at so active impact of fungi and bacterium mineralization of organic substances should be awaited.

Table 2. The amount of microorganisms arousing in soils while composting composite materials based on different polymers during 2 months, KOE/g of absolutely dry soil

Type of soils	Polymer base	Chapek medium (for fungi)	Chapek medium (bacterium)
№1	polyurethane	$1{,}12*10^6$	$1{,}88*10^7$
	CEVA	$5{,}00*10^6$	$1{,}55*10^7$
	Co-PA	$1{,}12*10^6$	$4{,}06*10^7$
	Lentex A4	$1{,}30*10^6$	$1{,}03*10^7$
	PVC	$1{,}85*10^6$	$2{,}15*10^7$
№2	polyurethane	$2{,}65*10^6$	$6{,}50*10^4$
	CEVA	$8{,}10*10^6$	$8{,}4*10^4$
	Co-PA	$3{,}83*10^6$	$2{,}7*10^5$
	Lentex A4	$6{,}49*10^6$	$1{,}79*10^5$
	PVC	$2{,}47*10^6$	$2{,}20*10^5$

Table 3. The amount of microorganisms arousing in soils while composting composite materials based on different polymers during 3,5 months, KOE/g of absolutely dry soil

Type of soils	Polymer base	Chapek medium (for fungi)	Chapek medium (for bacterium)
№1	polyurethane	$2{,}05*10^6$	$1{,}15*10^7$
	CEVA	$2{,}79*10^6$	$2{,}30*10^7$
	Co-PA	$8{,}67*10^5$	$2{,}88*10^7$
	Lentex A4	$9{,}82*10^5$	$3{,}95*10^6$
	PVC	$2{,}39*10^6$	$6{,}13*10^6$
№2	polyurethane	$3{,}76*10^6$	$1{,}38*10^6$
	CEVA	$4{,}87*10^6$	$1{,}09*10^6$
	Co-PA	$2{,}03*10^6$	$1{,}51*10^6$
	Lentex A4	$1{,}56*10^6$	$7{,}04*10^5$
	PVC	$5{,}17*10^6$	$1{,}24*10^6$

Table 4. The amount of microorganisms arousing in soils while composting composite materials based on different polymers during 5 months, KOE/g of absolutely dry soil

Type of soils	Polymer base	Chapek medium (for fungi)	Chapek medium (for bacterium)
№1	polyurethane	$5{,}56*10^5$	$5{,}11*10^8$
	CEVA	$4{,}30*10^5$	$4{,}92*10^8$
	Co-PA	$6{,}90*10^6$	$1{,}58*10^8$
	Lentex A4	$6{,}90*10^6$	$9{,}33*10^7$
	PVC	$6{,}90*10^5$	$3{,}76*10^7$
№2	polyurethane	$2{,}56*10^5$	$2{,}37*10^6$
	CEVA	$6{,}21*10^6$	$5{,}89*10^6$
	Co-PA	$3{,}28*10^6$	$5{,}85*10^7$
	Lentex A4	$9{,}00*10^5$	$2{,}66*10^7$
	PVC	$2{,}97*10^6$	$4{,}25*10^7$

At extension of composting time the tendencies remain: in the soil #1 with smaller amount of humus the total quantity of microorganisms are more than in the soil #2.

It should be mentioned that in spite of variety of fungi, *Mucor, Rhizopus* prevail in soil containing samples based on PBC, fungi of *Thrihoderma* genus prevail in soil with samples based on Lentex A4, fungi of *Penicillium* genus prevail in soil containing samples based on Co-PA.

In the tables 5-7 the results on quantitative and genus changes in bacterium on nitrite agar at different testing times of composting of composite materials are presented.

As it's shown in the table 5, the variety and quantity of bacterium in soil #1 is higher than in soil #2. *Nocardia* bacterium are absent in the soil enriched with humus. The quantity of bacterium of *Arthrobacter* genus is more in soils containing composite samples based on Co-PA, the quantity of bacterium of *Mycobacterium* genus is more in soils containing samples made of CEVA.

Mycobacterium and *Arthrobacter* fungi are characteristic for soils containing samples made of Co-PA, bacterium of *Mycobacterium* are characteristic for soils with samples made of PVC. Though *Micromonospora* bacterium are not typical for humus soils, but they well grow in soils with samples made of Co-PA and PVC.

The quantity of *Actinomyces* bacterium during the whole period of supervision doesn't change practically.

In the table #8 the results on quantity of arising CO_2 and CH_4 in different soils containing different composite samples are presented.

As it's shown from the table 8, we can not say that the intensity of respiration depends on the type of soil. Excreta of CH_4 takes place in the soil #1 depleted with humus after 3,5 months of composting. In the soil #2 excreta of methane was noticed only in case with PVC samples. In bare testing methane wasn't discovered.

The observed gas exchange is the sequence of breathing: of waste of grain production, of breathing of microorganisms and chemical oxidizing processes.

Table 5. The quantity of bacterium existing in soils while composting composite materials based on different polymers on nitrite agar during 2 months, KOE/g of absolutely dry soil

Type of soil	Polymer base	Myco-bacterium	Arthro-bacter	Nocardia	Micromo-nospora	Actino-myces	Total
№1	Polyure-thane	$8{,}14*10^8$	$5{,}53*10^8$		$5{,}60*10^7$	$7{,}50*10^7$	$1{,}11*10^9$
	CEVA	$6{,}27*10^9$	$1{,}24*10^8$	$2{,}00*10^7$	$2{,}00*10^7$	$1{,}04*10^8$	$6{,}55*10^9$
	Co-PA	$2{,}56*10^9$	$1{,}16*10^9$	$2{,}80*10^7$	$8{,}40*10^7$	$1{,}11*10^8$	$4{,}03*10^9$
	Lentex A4	$1{,}18*10^9$	$3{,}47*10^8$	$1{,}46*10^8$	$6{,}20*10^7$	$9{,}70*10^7$	$2{,}25*10^9$
	PVC	$1{,}99*10^9$	$5{,}14*10^8$	$7{,}30*10^7$	$2{,}40*10^7$	$2{,}82*10^8$	$2{,}57*10^9$
№2	polyurethane	$2{,}92*10^8$	$1{,}05*10^8$				$5{,}26*10^8$
	CEVA	$4{,}48*10^8$	$1{,}81*10^8$			$1{,}70*10^7$	$8{,}93*10^8$
	Co-PA	$1{,}04*10^8$	$7{,}46*10^8$		$5{,}83*10^6$	$1{,}75*10^7$	$1{,}42*10^9$
	Lentex A4	$1{,}82*10^8$	$6{,}76*10^8$			$2{,}00*10^7$	$1{,}02*10^9$
	PVC	$0{,}98*10^8$	$2{,}50*10^8$			$1{,}60*10^7$	$4{,}97*10^8$

Table 6. The quantity of bacterium existing in soils while composting composite materials based on different polymers on nitrite agar during 3,5 months, KOE/g of absolutely dry soil

Type of soil	Polymer base	Myco-bacterium	Arthro-bacter	Nocardia	Micromo-nospora	Actino-myces	Total
№1	Polyure-thane	$4,30*10^{8}$	$1,21*10^{7}$		$3,00*10^{7}$	$3,03*10^{7}$	$6,72*10^{8}$
	CEVA	$1,30*10^{9}$	$1,58*10^{8}$	$2,14*10^{7}$	$7,11*10^{7}$	$4,19*10^{7}$	$1,49*10^{9}$
	Co-PA	$5,64*10^{8}$	$1,65*10^{8}$		$5,14*10^{7}$	$2,92*10^{7}$	$7,77*10^{8}$
	Lentex A4	$1,03*10^{8}$	$7,52*10^{7}$		$1,54*10^{7}$	$7,53*10^{6}$	$2,06*10^{8}$
	PVC	$1,65*10^{8}$	$2,60*10^{7}$		$2,60*10^{7}$	$1,30*10^{7}$	$2,33*10^{8}$
№2	Polyure-thane	$1,16*10^{8}$	$1,35*10^{7}$	$4,44*10^{6}$		$4,44*10^{6}$	$1,54*10^{8}$
Type of soil	Polymer base	Myco-bacterium	Arthro-bacter	Nocardia	Micromo-nospora	Actino-myces	Total
	CEVA	$1,90*10^{8}$	$2,60*10^{7}$			$2,90*10^{6}$	$2,10*10^{8}$
	Co-PA	$3,87*10^{7}$	$2,29*10^{7}$		$3,28*10^{6}$		$6,62*10^{7}$
	Lentex A4	$4,85*10^{7}$	$1,58*10^{7}$		$5,37*10^{6}$		$8,02*10^{7}$
	PVC	$4,26*10^{7}$			$7,04*10^{6}$		$1,54*10^{8}$

Table 7. The quantity of bacterium existing in soils while composting composite materials based on different polymers on nitrite agar during 5 months, KOE/g of absolutely dry soil

Type of soil	Polymer base	Myco-bacterium	Arthro-bacter	Nocardia	Micromo-nospora	Actino-myces	Total
№1	Polyure-thane	$6,86*10^{7}$	$4,47*10^{7}$		$1,37*10^{7}$	$3,43*10^{6}$	$1,30*10^{8}$
	CEVA	$6,96*10^{7}$	$2,78*10^{7}$		$3,48*10^{6}$	$6,96* 0^{6}$	$1,04* 10^{8}$
	Co-PA	$1,34*10^{8}$	$8,23*10^{7}$	$1,03*10^{7}$	$5,14*10^{7}$	$1,03*10^{7}$	$2,78* 10^{8}$
	Lentex A4	$2,74*10^{7}$	$6,86*10^{6}$				$3,43* 10^{7}$
	PVC	$1,13*10^{8}$	$2,74*10^{7}$	$6,85*10^{6}$	$1,71*10^{7}$	$1,71*10^{7}$	$1,75* 10^{8}$
№2	Polyure-thane	$8,67*10^{8}$				$1,03*10^{7}$	$8,72* 10^{8}$
	CEVA	$1,39*10^{9}$	$9,28*10^{7}$	$2,32*10^{7}$		$2,32*10^{7}$	$1,53* 10^{9}$
	Co-PA	$2,22*10^{8}$	$1,71*10^{7}$		$1,71* 0^{7}$		$2,57* 10^{8}$
	Lentex A4	$2,77*10^{7}$	$3,69*10^{7}$	$9,23*10^{6}$		$7,39*10^{7}$	$1,57* 10^{8}$
	PVC	$7,11*10^{7}$	$1,78*10^{7}$		$8,89*10^{7}$	$3,56*10^{7}$	$2,31* 10^{8}$

According to [18], breathing of waste of grain production is approximately $1*10^{-2}$mg/g. Therefore, the decrease of the carbonic gas excreta can be explained both with depletion of organic filler as a result of microorganisms' action, table 2-7, Figure 1 and with the fact that biomass of fungi is much more than biomass of bacterium [19]. So, the decrease of quantity of fungi, table 2-4, leads to reduction of intensity of breathing.

During the experiment changes in chemical composition of soil were revealed. Depending on the type of soil and samples in them, the educed organic substances are different.

Soil with smaller content of humus and with composite materials based on polyurethane holds stearic acid. The other soil was unadulterated, though polyurethane samples without any filler had numerous holes, Figure 2, that are considered as the consequents of activity of microorganisms.

Table 8. The quantity of CO_2 and CH_4 excreting during composting of different composite samples in soils

Вид почвы	Polymer base	The quantity of CO_2 , мg/g of the sample			The quantity of CH_4 , мg/g of the sample		
		Testing time, months					
		2	3,5	5 * 10^{-2}	2	3,5 *10^{-6}	5
№1	Polyure-thane	2,3	2,1	1,30	отс	54,0	отс
	CEVA	0,9	0,35	0,68	отс	5,9	отс
	Co-PA	2,7	0,30	10,45	отс	17,3	отс
	Lentex A4	1,2	0,24	0,32	отс	6,2	отс
	PVC	2,3	0,40	3,04	отс	10,9	отс
№2	Polyure-thane	0,3	0,30	3,8	отс	отс	отс
	CEVA	1,0	0,59	2,67	отс	отс	отс
	Co-PA	3,7	0,27	10,67	отс	отс	отс
	Lentex A4	0,7	0,13	3,2	отс	отс	отс
	PVC	2,5	0,34	0,8	отс	9,1	отс

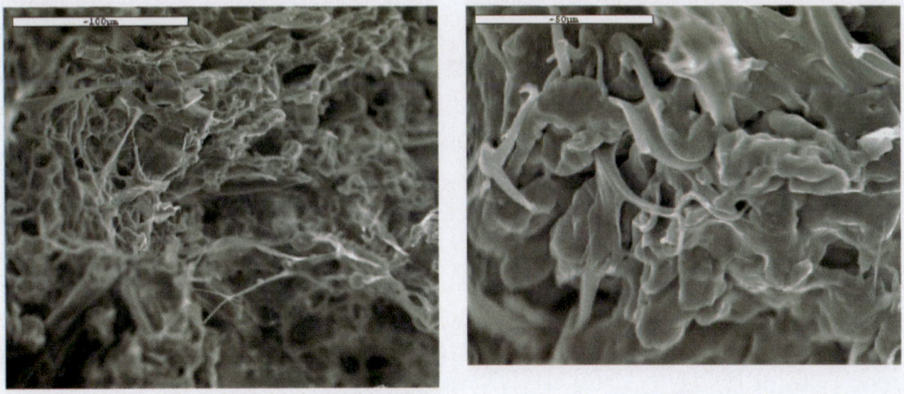

A B

Figure 1. Electronic microscopical picture of ship surface of the filled with organic and inorganic fillers polymer matrix based on Lentex A4: A) composted samples within 3,5 months; B) initial samples.

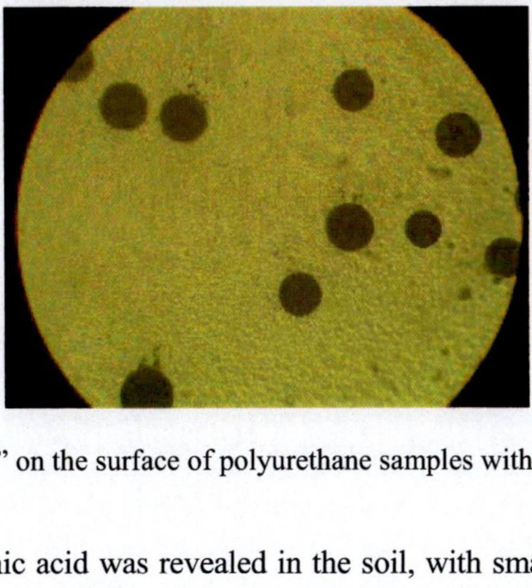

Figure 2. Photo of the “holes” on the surface of polyurethane samples without fillers being composted within 2 months.

Imide of dithiolcarbonic acid was revealed in the soil, with smaller content of humus and containing samples based on CEVA, after 3,5 months of storage while 1-nanodecene was present in the soil enriched with humus. “Behaviour” of soils with samples based on Co-PA differs to each other: there was pyrimidinetetraamine in the soil #1, and there was 1-isocyanododecane. Soils, containing samples based on Lentex A4 differ with the following: soil #1 keeps N,N-dimerhyldodecaneamid and 2-methyloximundecanon; soil #2 keeps 2-nonacosane. Allylmethylsulphide was founded out in the soil #1, containing samples on PVC, and imid of dithiolcarbonic acid was revealed in the soil #2. Concentration of organic substances is $1*10^{-3}$-$5*10^{-4}$ % and it enlarges from 1,5 to 3 times with time duration. It has to be marked that the singled out substances are not toxic to the environment [20].

Thus, the change of dominating groups of microorganisms on the surface of composite materials in the process of their composting was revealed in the result of the carried out investigation. The quantity of fungi diminishes in 10 times and the quantity of bacteria enlarges as much times and dominates. The lowering of fungi quantity is connected with exhausting of easy to access of organic and inorganic fillers, being part of composite materials. The growth of bacteria quantity is connected with the beginning of biological destruction and mineralization o f polymer base.

REFERENCE

[1] Wolfram Tanzer. *Biologisch abbaubare Polymere*. Deutscher Verlag fur Grunstoffindustrie. Stuttgart, 2000, 200s.

[2] O.A.Legonkova, L.A. Sukhareva. *Thousand and one polymers*. UPGroup, Moscow, 2004, 270 p.

[3] O.A.Legonkova, O.A.Sdobnikova, A.L.Peshehonova, A.A.Bokarev. *Biodegradable polymers in packaging technology*. «Tara I upackovka», 2003, №6.

[4] O.V.Suchugova, N.N. Kolesnikova, A.N.Lihachev, A.A.Popov. *Plasticheskie massi*, 2004, №9, pp. 29-32.

[5] Patent № 2257045, RF.

[6] E.Z. Tenner, V.K.Shilnikova, G.I.Pereverzeva. *Pracice for microbiology*, M., Koloss, 1979, 215 p.

[7] E.L.Pehtysheva. *Biodamages and defence of nonfoods*. Moscow, Masterstvo, 2002, 224 p.

[8] L.A. Sukhareva, M.I. Gubanova, O.A.Legonkova. *Biochemically stable powder compositions in food industry*. Moscow, VNIIMP, 2002, p.177.

[9] V.S. Jakovlev, E.V.Bakirova, N.G.Chulkov. Investigaion of inner surface of volumes for storage of ethyl alcohol. Conf. “*Storage of food products and raw materials*”, Moscow, 1999, pp. 53-54.

[10] P.I.Zubov, L.A. Sukhareva. *Structure and properties of polymer coverings*. Moscow, Chemistry, 1982, 256 p.

[11] L.A. Sukhareva and etc. Physical and chemical basis for modification of defensive polymer coverings. *Физико-химические основы модификации защитных полимерных покрытий*. Kotovitzhi, 1998, pp. 81-86.

[12] L.A. Sukhareva and etc. The ways of advancement of durability of polymer coverings based on creation of well-ordered structure. Conf. “*New chemical technologies for production and use*”, Penza, , 2000, p.60.

[13] L.A. Sukhareva and etc. Biochemical stable defensive coverings for water-supply. Conf. "*Durability and protection of construction from corrosion*", 1999, pp. 577-579.
[14] L.A. Sukhareva and etc. Biostable coverings of large capacities. Conf. "*Ecological problems of biodegradation of industrial, building materials and waste of manufacture*", Penza, 2000, p.45.
[15] GOST 9.049-91. Unified system from corrosion and aging. *Polymers and their components*. Methods of laboratory testing on stability to fungi impact.
[16] G.Shlegel. *Microbiology*, Moscow, Mir, 1972, 473p.
[17] E.N.Mishustin. *Ecological geographical changing of soil bacteria*. Leningrad, 1947, 325 p.
[18] L.A. Trisvjatskii. *Grain storage.*Moscow, Agropromizdat, 1986, 343 p.
[19] D.G.Zvjagintzev. *Soil and microorganisms*. Moscow, 1987, 200 p.
[20] I.I.Lazarev. *Harmful substances in industry*. Moscow.

In: Preparation and Properties of Monomers, Polymers… ISBN 978-1-60021-557-5
Eds: A. Ballada and G. E. Zaikov, pp. 175-198

Chapter 19

NEW TECHNOLOGIES FOR FAST LIQUID-PHASE CHEMICAL PROCESSES

V. P. Zakharov[1], Al. Al. Berlin[2] and G. E. Zaikov[3]

[1]Bashkirian State University, 32, Frunze st., Ufa, 450074 Bashkortostan, Russia
[2]N.N. Semenov Institute of Chemical Physics, Russian Academy of Sciences, 4, Kosygina st., Moscow, 119991, Russia
[3]N.M. Emanuel Institute of Biochemical Physics, Russian Academy of Sciences, 4, Kosygina st., Moscow, 119991, Russia

ABSTRACT

Change of geometry (design) of mixing zone of turbulent flows limited by impenetrable wall, physical parameters and reagents dynamics allow easy control of processes of mass and energy transfer in accordance with specificity of proceeding process. Physico-chemical basis of fast processes realization in turbulent regime serves as a base for creation of a number of energy- and resource-saving high-productive technologies on the base of novel type of reactors of applied chemistry – tubular apparatus of jet type.

Keyword/Phrase index: fast chemical reaction, plug-flow tubular turbulent apparatus, ethylene-propylene rubbers production, chlorobutyl rubber, low-molecular products production.

INTRODUCTION

Various processes in which reagents in consequence of interaction undergo deep conversions followed by changing of internal structure, substances composition and aggregate state are realized in chemistry and applied chemistry. Chemical-engineering processes use to contain the physical ones also, in particular liquids movement, their mixing, heating and cooling, dispersion, etc. It is worth saying that technology of various chemical products production as a rule comprises chemical and physical processes of the same type characterized by common regularities. But for all that processes of different productions are carried out in apparatus of analogous principles of operation. According to the classification of applying chemistry apparatus there are only two theoretical (limit) models of instruments: apparatus of ideal displacement and of ideal mixing. Real ever-functioning reactors are representing apparatus of intermediate type with some approximation to apparatus of mixing or displacement. It is because particles residence time in them distributes more uniformly than in ideal displacement apparatus but never becomes level as in ideal mixing apparatus.

Intensification of chemical processes, reducing of their steel intensity, energy- and resource economy, creation of scientifically well-founded, economical, compact technologies of high ecological safety and apparatus for them are the priority directions of modern technology development. It is obvious that for such problem-solving routine investigations of processes and apparatus in production quantities and conditions are usually complex and prolonged. That is why modelling is very significant; it allows studying regularities of processes proceeding and apparatus functioning with the help of models under conditions permitting application of resulting data independently from production quantities.

It is possible to assume that kinetic and hydrodynamic methods of reactors analysis and design are well advanced at present. Methods of computer simulation and modelling are widely used. So, we can say that if we know processes kinetic and hydrodynamic parameters and fundamental particularities of reactor functioning we can calculate all process characteristics and its structure, we also can predict effectiveness of apparatus operation and consumer properties of chemical production. Meanwhile criterion function development for calculation and organization of novel processes and optimization of present productions require overcoming of a big number of problems in production practice. This principle is satisfactory enough for processes with low or medium reactions rates, when creation of isothermal conditions in apparatus is easy. In this case it is easy to calculate and reproduce in working conditions all characteristics of chemical process and to control the last ones.

We may see the inverse picture under fast processes proceeding with high rates – characteristic duration of chemical reaction is less than 0,001-0,1 seconds. This class of reactions was not studied well enough until recent times, as it turned out that it comprised a large number of important chemical and mass-transfer physical processes realized in industry. When carrying out chemical reactions proceeding with high rates (characteristic time of reaction is less than time of transfer) complex temperature and concentration fields are formed in the reaction zone. Fast chemical processes proceed in diffusion field as a rule and they are limited by reagents inlet into reaction zone. So, special approach to the technological scheme apparatus design is needed in every concrete situation. Modern productions processes with application of standard stirred tank reactors are ill-controlled, proceed under conditions wide of the isothermal. They also characterized by unwarranted long times of reagents residence in reaction zone, noticeable reduction of final product total yield, high content of side substances, deteriorated heat removal, etc. When one synthesizes polymers in standard stirred tank reactors undesirable change of their molecular structure, molecular mass (MM), molecular-mass distribution, formation of heterogeneous products and consequently deterioration are observed. Uncontrolled reaction temperature regime may lead to sudden effects, in particular strong hydro- and pneumatic strokes, rejections, heat explosions, etc. That is why selected for fast processes in liquid phase and widely used in industry tank reactors of continuous action ($V = 2 \div 30\ m^3$) with intensive mixing and developed heat removing system never can be found successful and optimal.

The use of ideal plug-flow reactors of continuous action for the same aims is even grater unsuccessful because even under relatively slow chemical reactions realization of isothermal regime in reactors of such type is impossible. Plug-flow reactors are usually function in adiabatic or intermediate but not isothermal (including external heat removal) regimes. So, we may conclude, that practically all reactors using in industry for fast processes are non-optimally adjusted and generally ineffective in production. Consequently the quality of resulted polymers are not optimal and processes are engineering, socially and economically imperfect (reduction of resulted product yield, deterioration of its quality, high content of non utilizable waste, unjustified high consumption of raw materials, electric power, etc.).

Thus, under realization of quite fast chemical processes in real productions it is necessary to take into account that true chemical mechanism of proceeding processes is complicated by totality of physical processes such us diffusion, heat- and mass-transfer, hydrodynamics. Fast processes proceed in diffusion region and require macro-kinetic approach to a solution of this problem.

For the first time deep investigations of ultra-speed liquid-phase processes began to develop by the example of isobutylene oligomerization and polymerization as a classical model reaction. Sufficiently great number of fast chemical and mass-transfer physical processes was studied then. Necessity of decrease of reaction apparatus outline dimensions down to sizes commensurable with "torch" sizes proved to be a very important scientific and applied conclusion. This lead to the development of compact highly productive tubular turbulent apparatus functioning in a novel quasi-ideal plug-flow regime. Novel technology of fast chemical reactions realization in turbulent regime is also effective under many mass-transfer processes (extraction, mixing, mixing of flows with various viscosity and density, dispersion, saturation of liquid mediums by gases, etc.). Application of tubular turbulent apparatus allows increasing effectiveness of external heat removal that simultaneously determines possibility of their use as a compact heat exchangers.

Accordingly to aforesaid fundamental regularities of fast liquid-phase chemical, heat- and mass-transfer processes in turbulent flows are considered in the book. Novel energy- and resource-saving technologies on the base of compact tubular turbulent apparatus are presented.

PLUG-FLOW TUBULAR TURBULENT APPARATUS – THE NOVEL TYPE OF INDUSTRIAL APPARATUS

Novel apparatus types for effective fast chemical reactions realization in turbulent regime were worked out (Figure 1, Table 1). For liquid-phase processes methods for calculation and construction of high-productive compact tubular turbulent apparatus of plug-flow of jet type but working in turbulent flows were developed. Apparatus combine the best technical and technological advantages that are known in industry of mixing and plug-flows reactors and are differ by the presence of their own original inhering only to them fundamental particularities. This type of apparatus is not considered in active classification of chemical technology apparatus [1-3].

Four modifications of tubular turbulent apparatus irreplaceable under realization of fast chemical processes and suitable for carrying out of mass-exchange liquid-phase processes in industry were developed: cylindrical (with constant reaction zone diameter along apparatus length); shell-and-tube (pipe bundle washing by coolant); "zone" (concatenate independent adiabatic reaction zones separated by cooling zones) and divergent-convergent (with several concatenate sections with local hydrodynamic resistances) designs principally differing from known industrial apparatus of mixing and plug-flow, by this tubular turbulent apparatus may function in both regimes plug-flow and mixing.

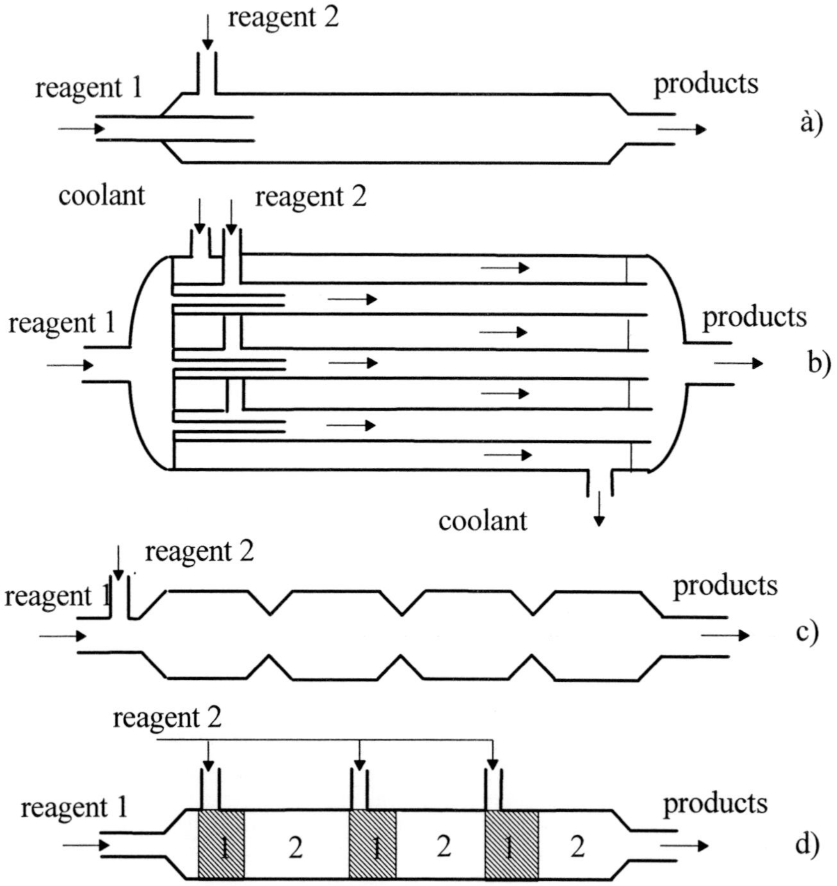

Figure 1. Tubular turbulent apparatus of cylindrical (a), divergent-convergent (b), shell-and-tube (c) and "zone" (d) construction (1 – reaction zone, 2 – cooling zone).

Relative disadvantages of compact tubular turbulent apparatus of the first generation are: fast decay of turbulent diffusion coefficient $\mathbf{D_t}$ in reaction zone and along its length (over 1-2 calibres) and low effectiveness of external heat removal. In a number of cases they succeeded in intensification of external heat removal. Shell-and-tube turbulent reactors were worked out with this aim. Using pipe bundle washing by coolant and maintaining required productivity one may significantly improve external heat removal especially under process carrying out in quasi-isothermal conditions. If under realization of fast chemical processes on the base of two constructions of turbulent apparatus effective heat removal can not be provided then for the avoidance of inadmissible overheating of reaction mixture one may use "zone" model. The reagents (or catalyst) portion that determines admissible temperature rise in reaction zone is introduced into every zone, where reaction proceeds. The bigger the number of reaction zones, the bigger the possibilities of fast chemical reactions control.

Table 1. Indexes* of reactors work of continuous action under fast chemical reactions

Indexes		Tubular turbulent apparatus	Plug-flow reactors	Stirred reactors	Two-stage stirred reactor (cascade)
Productivity		1	0,1	0,1-0,3	0,5-0,6
Specific productivity		1	0,001	0,001-0,002	0,004-0,005
Steel intensity		1	100	1000	600
Power intensity		1	1	12	12
Residence time		1	500	350	100
Time of regime establishment (after shutoff)		1	50	400	300
Conversion degree for the time τ_r		1	1	0,75±0,05	up to 0,9
Reactor volume	at n = 1	1 (1)	1000 1000	4000 21500	
at β = 0,9 and β = 0,99	at n = 2	1 (1)	1000 1000	10000 100000	3000 8000
Characteristics of polymer formed under fast polymerization reaction:					
MMD		2,5±0,5	10±1	5±1	3±1
M_n		corresponds to calculation	lower than calculated	lower than calculated	lower than calculated
M_w		-//-	-//-	-//-	-//-
M_z		-//-	-//-	-//-	-//-
Specific charge on:					
catalyst		1	1	1,5-2	1,3
water		0,8	0,8	1	0,9
raw materials		0,75	0,75	1	0,85

* Values of indexes are given in relation to tubular turbulent reactor.

Tubular turbulent apparatus of the second order of divergent-convergent design that essentially by all positions surpass any known apparatus of chemical technology are the best in their technical characteristics. In particular, their distinctive features are: possibility of receipt of required turbulence level and consequently intensity of mixing and heat-transfer, maintaining of practically constant turbulence level in apparatus volume and along its length at significantly lower flow rate; possibility of formation of self-similar regime in relation to viscosity and criterion Reynolds Re etc.. As a consequence apparatus of this type are irreplaceable for work with any mediums including high-viscosity polymer solutions and allow increasing of process productivity; high effectiveness under work with multi-phase systems because there is fine phases dispersion with receipt of homogeneous emulsions and foams due to geometry of divergent-convergent canal, significantly higher (in order and higher) solution rate and gas amount in liquid in comparison with barbotage or intensive mechanical mixing; increase (in 2-4 times) of coefficient of heat-transfer through the outside wall in comparison with apparatus of cylindrical design. Fast process proceed in optimal regime if the ratio $\tau_{ch} > \tau_{turb} > \tau_{meso} > \tau_{micro}$ is met, by this decrease of reaction zone radius R leads to reduction of characteristic mixing times that represents the key to realization of fast processes in optimal conditions. Change of liquid flows rate in apparatus of divergent-convergent design and ratio diameter diffuser and confuser d_d / d_c is the only but effective way of influence on the dispersion character and quality of resulted emulsions. Decrease of d_d and increase of turbulence level in reaction volume (at the expense of V, D_t and ε increase) promote receipt of fine systems. Apparatus of divergent-convergent design may function in both plug-flow and mixing regimes that is reached firstly by the change of apparatus geometry and reaction mixture movement rate. Formation of circulation zones provides flows in apparatus of this type. Under movement of multi-phase mediums under the action of centrifugal forces the separation effect may be observed leading to the appearance of heterogeneities in components distribution in disperse system with phase separation in dependence of density of participating in process reagents.

Thus, tubular turbulent apparatus of any construction under formation in them of quasi-plug-flow mode in turbulent flows represent principally novel type of industrial compact high-productive apparatus combining the best technical and technological advantages of jet stirred and plug-flow reactors using in industry at present and having their own original particularities.

Realization of tubular turbulent apparatus under fast processes carrying out at conditions of industrial production allows creation of continuous energy- and resource-saving technologies.

POLYMER PRODUCTION

Ethylene-Propylene Rubbers Production

The scheme of continuous solution copolymerization of ethylene and propylene (Figure 2) was developed and introduced into production, tubular turbulent apparatus are used in it at the following stages:

1) preparation of homogeneous gas-liquid mixture and its introduction into parallel working polymerizers (Figure 2, positions 2-4, Figure 3, positions 8, 9, 10, 12);
2) formation of sites of macromolecules propagation (Figure 2, position 6). (Patent of Russia No.2174128; priority of 27.03.2000, published on 27.09.2001, bulletin 27).

Under formation of homogeneous gas-liquid mixture (Figure 3) cooled components in definite proportions under pressure are introduced into comb (position 7) and tubular turbulent apparatus of divergent-convergent design (positions 8 and 9) in which developed turbulent regime of flow and consequently preliminary mixing are provided.

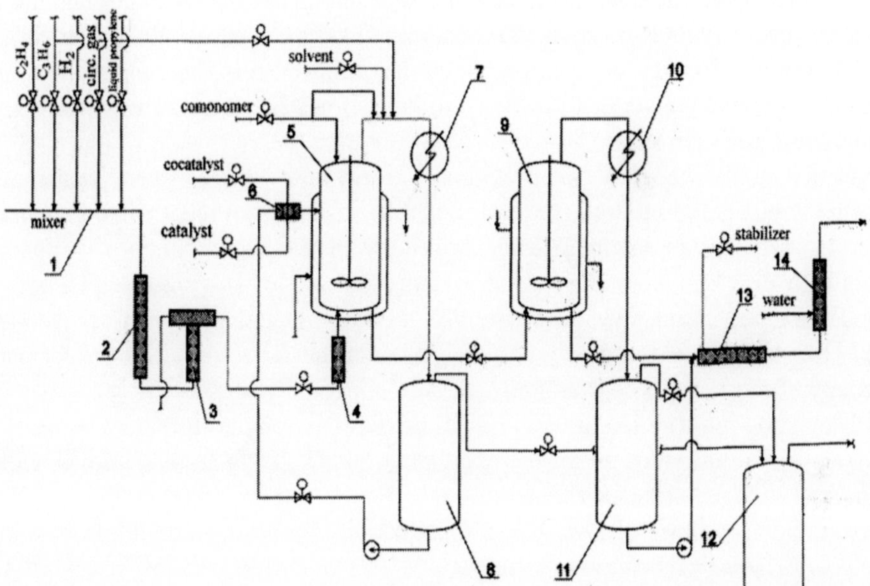

Figure 2. Principle scheme of SKEP(T) production. 1 – comb of preliminary mixing; 2, 3, 4, 6, 13, 14 – tubular turbulent apparatus; 5, 9 – polymerizers; 7, 10 – condenser of blowing back; 8, 11 – collector condenser blowing backs; 12 – separator-eliminator of liquid phase of blowing backs.

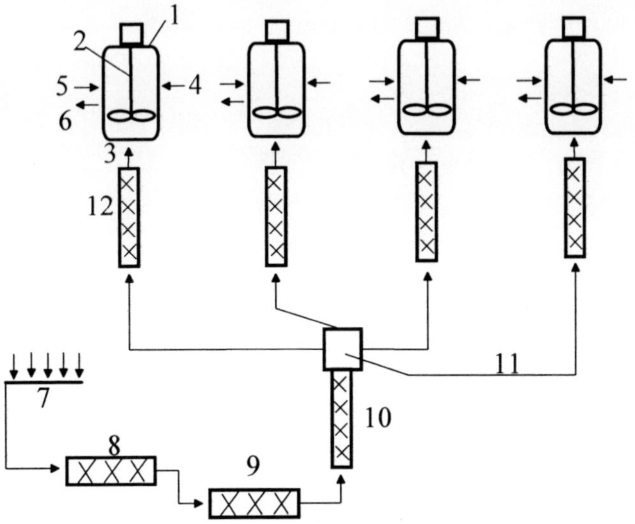

Figure 3. The method of continuous solution copolymerization and reactor-distributor for it realization. 1 – polymerizer; 2 – drive of mixing device; 3 – connection pipe for introduction of gas-liquid mixture; 4, 5 – connection pipes for separate introduction of catalytic complex components; 6 – connection pipe for copolymer taking off; 7 – comb; 8, 9, 12 – tubular turbulent apparatus of divergent-convergent design; 10 – tubular turbulent apparatus-distributor; 11 – transfer tubes; 3 – connection pipes for introduction of gas-liquid mixture into polymerizer.

Gas-liquid mixture again mixers at conditions of turbulent flow when getting into tubular turbulent reactor-distributor (position 10). Gas-liquid mixture is introduced into transfer tubes (position 11) practically does not change conditions of movement. Then, via tubular turbulent apparatus of divergent-convergent design (position 12) gas-liquid mixture is introduced into parallel working polymerizers.

Developed turbulent flow at preliminary stage provides uniform distribution of components in gas-liquid mixture (position 8 and 9). Tubular turbulent apparatus-distributor (position 10) provides maintaining of homogeneity of gas-liquid mixture components in volume at their distribution among transfer tubes (position 11). This forms equal structure of gas-liquid mixture introducing into parallel working reactors-polymerizers. Gas-liquid mixture before introduction into polymerizer is again subjected to turbulization for providing homogeneity of components distribution.

Stage-by-stage preparation of gas-liquid mixture in tubular turbulent apparatus and distribution of its introduction to the parallel working polymerizers allows receipt of copolymers with equal properties in various reactors and at other equal conditions of polymerization the increase of total process productivity was observed and productivities of separate polymerizes by this became identical.

At the stage of formation of sites of macromolecules propagation under ethylene with propylene copolymerization initiated by V-Al catalytic systems (separation of fast stage of active sites formation and initiation and slow stage of copolymerization itself) in novel technological scheme of SKEPT production tubular turbulent apparatus is installed (Figure 2, position 6, Figure 4).

In this case catalyst is introduced via nipple 1 by inside canal 2 of pre-reactor (Figure 4) and solvent via nipple 3. The flow after passing via turbulization section 4 equalizes catalyst concentration over flow volume and introduces into 5. By inside canal the flow comes to the divergent-convergent section 6 where its turbulization occurs and then to the nozzle 7, where flow additionally acquires non-stable hydrodynamic current. At output of nozzle the flow reaches Veber's criterion 0,5 and as a result of interaction with gas medium begins to break but not earlier than at 50 mm from nozzle head.

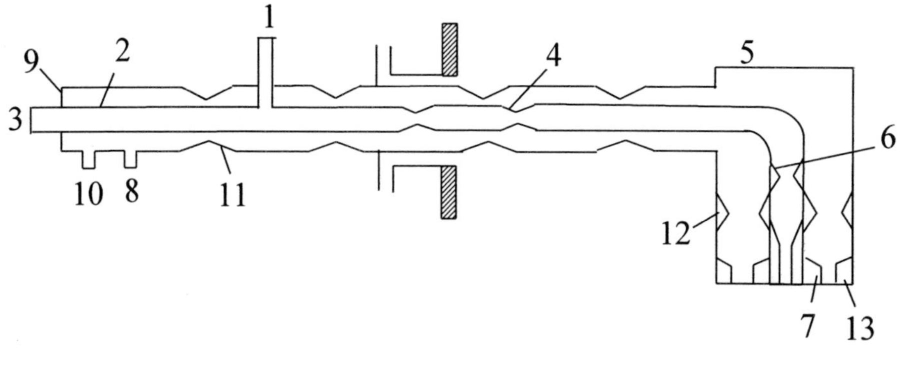

Figure 4. Tubular turbulent apparatus for formation of homogeneous high-dispersed sites of macromolecules propagation under ethylene and propylene copolymerization.

Cocatalyst is introduced via socket 8 into outside canal 9 in which via socket 10 the solvent is preliminary introduced. The flow passing turbulization section 11 equalizes cocatalyst concentration over volume and is introduced into outside canal 5. In outside canal the flow is also preliminary turbulized in divergent-convergent section 12 and then via nozzles 13 is introduced into gas medium. Getting out from head of nozzle 13 the flow has value of Veber's criterion 0,5 and as a result of interaction with gas medium is broken. Breaking process should start not earlier than at 50 mm from nozzle head.

Fast dilution of catalytic complex components is provided by turbulization of mixing flows that is realized by configuration of inside surface of each canal of tubular turbulent pre-reactor. By this the flow by turns undergoes pressing and broadening promoting formation of turbulences that provide full and fast mixing of liquids with various density over the whole volume of moving flow. Mixing efficiency is provided in at least two divergent-convergent sections.

Under catalytic complex components mixing fast reaction of V-Al catalyst formation proceeds and growing crystal active sites of high activity is as a rule precipitated onto metallic surface of device applying for components mixing. Gradually the thickness of deposited layer is increased, copolymerization reaction is additionally proceeded on it and polymer layer is formed that leads to driving of holes of catalytic complex solutions introduction. Moreover, formed at these conditions catalytic complex due to the absence of steric difficulties forms large crystals with perfect structure covering the most part of active sites inside of the crystal. That is why creation of conditions of separate introduction of catalyst components in tubular turbulent pre-reactor of developed construction (Fig. 4) excluding the deposition of crystals on apparatus surface. Moreover, creation of conditions of flows interaction mainly in gas phase over reaction mass surface on the one hand creates conditions of catalytic complex components fast mixing with formation of small catalyst particles, on the other hand – formation of these crystals occurs in gas medium containing large amount of gaseous monomers which are precipitated on active sites of growing crystal faces preventing from formation of perfect and large crystals.

Results of carrying out of stage of V-Al catalytic system active sites formation in turbulent regime are presented in Table 2.

Table 2. Formation of active sited in turbulent regime under ethylene and propylene copolymerization initiated by $VOCl_3$-$Al(C_2H_5)_2Cl$ catalytic system

No	Copolymer grade	Time interval between experiments, h	Catalytic complex consumption per ton of copolymer	Copolymer yield, kg/hour	Mooney viscosity
1	SKEP	0	3,0	920	48-50±1
2		24	3,2	950	48-50±1
3	SKEPT with DCPD	0	2,9	1020	50-52±1
4		24	3,1	1010	50-52±1
5		48	2,9	1020	50-52±1
6	SKEP	0	3,5	800	45-47±1
7		24	3,5	750	43-45±1
8		48	3,5	780	45-47±1

1-5 – formation of active sites in turbulent regime with use of tubular turbulent pre-reactor (Figure 4)
6-7 – separate introduction of catalytic system components directly into tank reactor-polymerizer.

Preparation and introduction into polymerizer of catalytic complex components in tubular turbulent pre-reactor (Figure 4) provides production of copolymer (SKEP and SKEPT) with good and stable in time properties (Table 2). Analogously to laboratory results received under separation of stage of formation of active sites and initiation of isoprene polymerization on Ti-Al catalytic system at conditions of industrial production under ethylene of propylene copolymerization increase of copolymer yield and its molecular mass and also decrease of catalyst consumption are observed.

Chlorobutyl Rubber Production

On the base of studying of specificity of process of butyl rubber chlorination in solution by molecular chlorine, conditions of formation of self-similar field of liquid flows and regularities of multi-phase flows in tubular canals compact tubular turbulent pre-reactor-chlorator (Figure 5) with original nozzles (Figure 6) (Patent Russia No. 2170237, priority from 30.07.1999, published on 10.07.2001, bulletin No. 19; Patent Russia No. 2170238, priority from 30.07.1999, published on 10.07.2001, bulletin No.19).

Nozzles are placed in series along tubular reactor and have on the outside surface the flutes forming with inside surface of box wall screw canal (Figure 6), and inside surface of nozzles has on flanks conical lugs with diameter in 1,3-3 times lower than in cylindrical part of inside cavity of nozzles and forming tubular turbulent reactor of divergent-convergent design (Figure 5). Screw canal and inside cavity of nozzle are connected by holes and/or cuts, and beyond last nozzle in the course of gas-liquid flow one or more static mixers are placed.

Forming in tubular turbulent reactor of divergent-convergent design gas-liquid mixture is separated in placed in series along the flow nozzles into inter-communicating flows that are united into common flow at the output of last nozzle. Taking into account that the volume of gas medium in process exceed the volume of BR solution after preliminary mixing of gas-liquid mixture one should separate the flow into at least two parts – it is divided in the nozzle into peripheral and axial flows that provide creation of optimal conditions for chlorination reactions proceeding.

In production of chlorobutyl rubber it is advisable to use tubular turbulent apparatus of cylindrical or divergent-convergent designs also at other stages of technological process, in particular under neutralization of resulted chlorobutyl rubber solution, scouring of chlorobutyl rubber solution by water from salts and other compounds, washing of inverse solvent; under introduction into chlorobutyl rubber solution of stabilizer and anti-agglomerator and also instead of all intensive including in-volume agitators with mechanical mixers (Figure 7).

Thus, principally novel, unique, economic, continuous process of chlorobutyl rubber production with the use of tubular turbulent apparatus of original construction operating in regime of high flows turbulence on at least four stages of technological scheme was worked out (Figure 7). In novel process stirred tank reactors are substituted by tubular reactors at stages of butyl rubber chlorination (position 4), neutralization (position 5), stabilizer and anti-agglomerator introduction (position 12, 15). It is possible to replace by tubular apparatus the flush column also (position 9) where water washing of solvent is carried out.

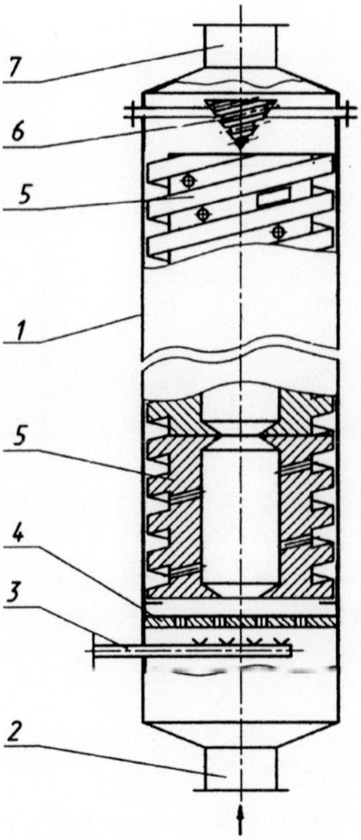

Figure 5. General form of tubular turbulent reactor of divergent-convergent design with nozzles for chlorination of BR. 1 – apparatus box; 2 – connecting pipe for introduction of butyl rubber solution; 3 – connecting pipe of gas mixture or gas introduction; 4 – flow distributor; 5 – nozzles; 6 – static mixer; 7 – connecting pipe for gas-liquid mixture taking off.

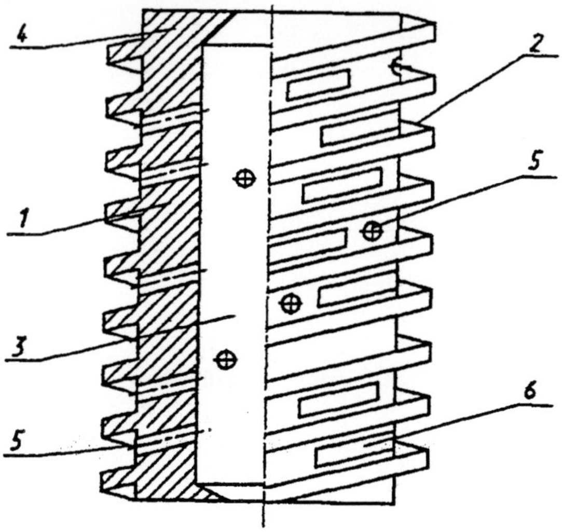

Figure 6. The scheme of nozzle in tubular turbulent reactor. 1 – hollow box; 2 – screw thread; 3 – inside cavity; 4 – conic lugs (convergents); 5 – holes; 6 – cuts.

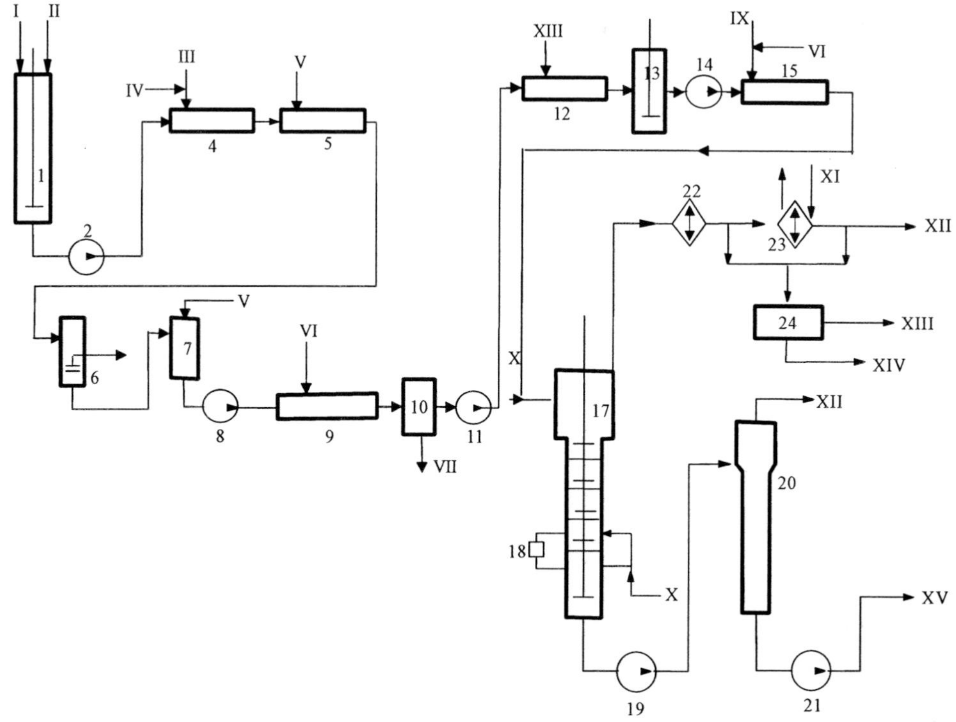

Figure 7. Technological scheme of chlorobutyl rubber production with the use of tubular turbulent apparatus of jet type: 1 – apparatus of chlorobutyl rubber solution preparation; 2, 8, 11, 14, 19, 21 – pumps; 12, 15 – tubular turbulent apparatus for introduction into chlorobutyl rubber of stabilizer-antioxidant and anti-agglomerator; 4 –tubular turbulent apparatus-chlorator; 5 – tubular apparatus-neutralizer; 6 – filter; 7 – collector; 9 – tubular turbulent apparatus for water washing of solvent; 10, 24 – sediment boxes; 13 – neutralizer; 17 – degasifier; 18 – orifice device; 20 – vacuum degasifier; 22, 23 – condenser. Flows: I – solvent; II – butyl rubber crumb; III – chlorine; IV – nitrogen; V – alkali water solution; VI – water; VII – washing water for purification; VIII – stabilizer-antioxidant solution; IX – anti-agglomerator suspension; X – vapor; XI – coolant; XII – to the vacuum line; XIII – benzene dewatering; XIV – water for stripping of organic compounds; XV – pulp into concentrator.

The process is remarkable for compactness and guarantees significant energy- and resource-saving, high environmental safety, easy control, process productivity increase (in hundreds of times), significant decrease of production space, increase of products quality. Technology of chlorobutyl rubber production is significantly simplified, investment expenses and products prime cost are decreased. Process is continuous.

Liquid Oligo-Piperylene Rubber Production

Traditional technological scheme of liquid oligo-piperylene rubber of SKOP sort production [4,5] independently on catalytic system activity includes as main apparatus stirred tank reactor (12 m^3) or cascade of them (in 1,5 m^3) equipped by mechanical mixers and also by outside and inside coolers (coolant is inverse water) (Figure 8a).

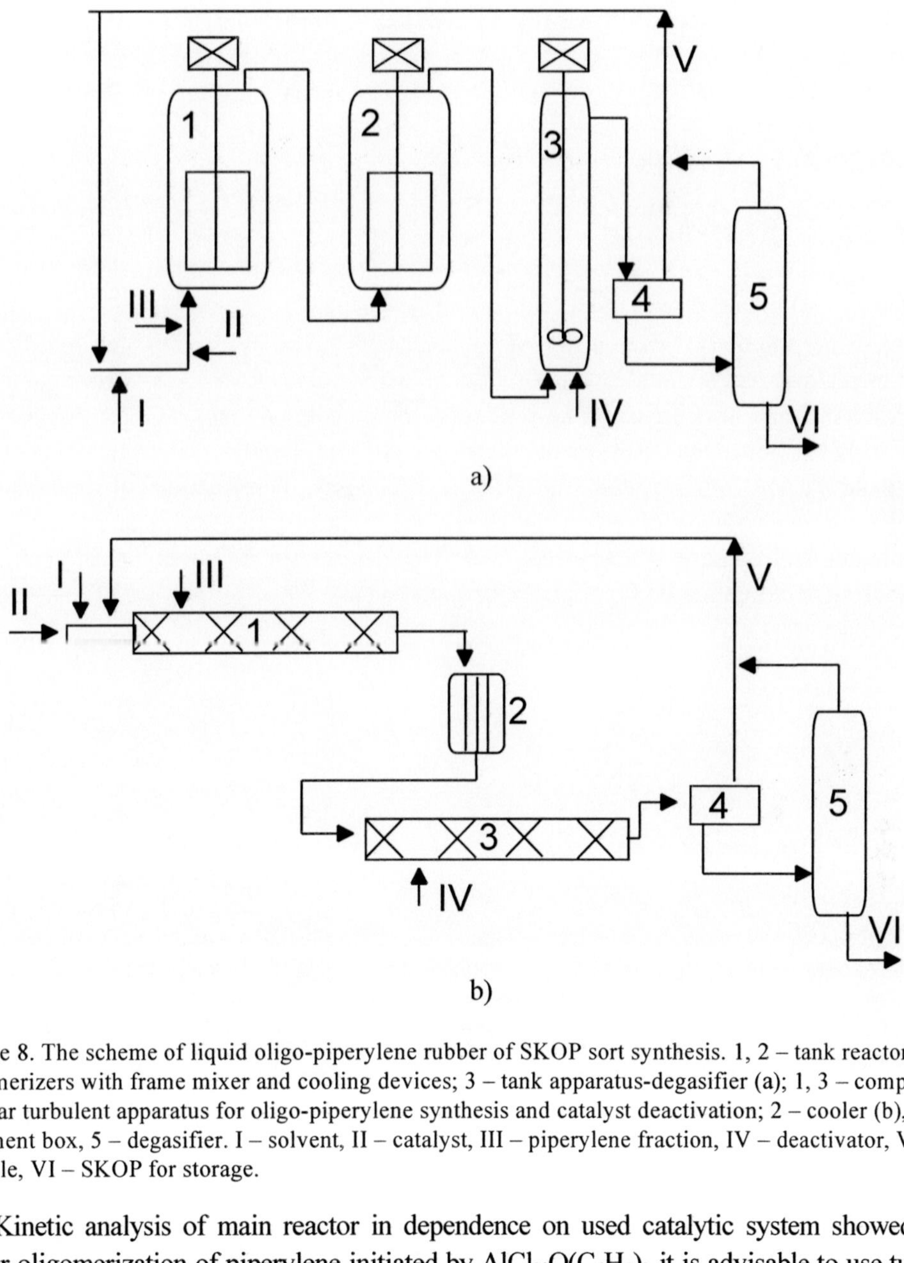

Figure 8. The scheme of liquid oligo-piperylene rubber of SKOP sort synthesis. 1, 2 – tank reactors-polymerizers with frame mixer and cooling devices; 3 – tank apparatus-degasifier (a); 1, 3 – compact tubular turbulent apparatus for oligo-piperylene synthesis and catalyst deactivation; 2 – cooler (b), 4 – sediment box, 5 – degasifier. I – solvent, II – catalyst, III – piperylene fraction, IV – deactivator, V – recycle, VI – SKOP for storage.

Kinetic analysis of main reactor in dependence on used catalytic system showed that under oligomerization of piperylene initiated by $AlCl_3 \cdot O(C_6H_5)_2$ it is advisable to use tubular turbulent apparatus. Not high values of dynamic viscosity of liquid oligo-piperylene rubber at output of reactor (at 283 K, $\mu = 1$ mPa·sec) allows also perfecting of stage of catalyst deactivation by propylene oxide or water (Figure 8b).

Novel technology of piperylene oligomerization (Figure 8b) guarantees saving of main principles of working technological schemes, increase of specific productivity of main reactor, reduce of specific weight of side reactions and secondary process proceeding at the expense of significant reduce of τ_r.

Anti-Agglomerator Production for Synthetic Rubbers

At present calcium stearate (anti-agglomerator) is produced by two-stage method:

$$C_{17}H_{35}COOH + KOH \rightarrow C_{17}H_{35}COOK + H_2O$$

$$2C_{17}H_{35}COOK + CaCl_2 \rightarrow (C_{17}H_{35}COO)_2Ca + 2KCl$$

The second stage – interaction of potassium stearate (water solution) with calcium chloride (water solution) is the reaction of ion exchange and proceeds with high speed, i.e. in diffusion regime that required intensification of mass-exchange process. Received in this work results on fine suspension synthesis allow recommending of application at this stage the tubular turbulent apparatus of divergent-convergent design (Figure 9) consisting of cascade of three concatenated mini-reactors of jet-type. Proposed construction of apparatus is characterized by high value of turbulence kinetic energy density dissipation and consequently by minimum micro-mixing characteristic time. This determines the possibility of creation of local shifting deformations and receipt of fine suspensions.

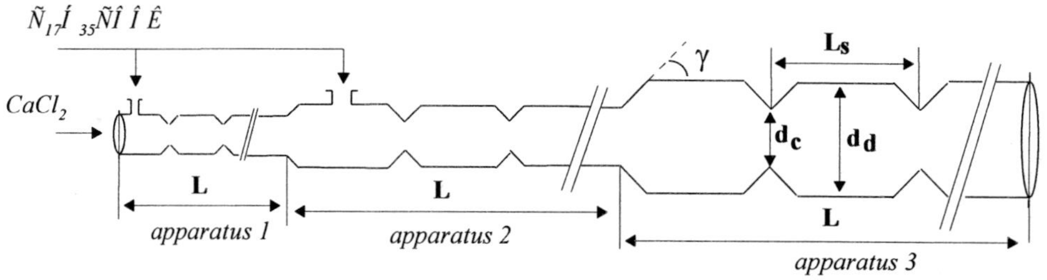

Figure 9. General view of tubular turbulent apparatus for stage of interaction of potassium stearate with calcium chloride. Apparatus 1: d_d = 30 mm, d_c = 20 mm, L_s = 90 mm, L = 0,9 m, $\gamma = 35 \pm 5^0$, sections number 10; apparatus 2: d_d = 40 mm, d_c = 20 mm, L_s = 120 mm, L = 1,4 m, $\gamma = 35 \pm 5^0$, sections number 12; apparatus 3: d_d = 50 mm, d_c = 25 mm, L_s = 150 mm, L = 3 m, $\gamma = 35 \pm 5^0$, sections number 20. Total apparatus length is 5,5 m.

Every section of turbulent reactor is calculated upon optimal operation at various process productivity: apparatus 1 – up to 2,5 m^3/hour of summary liquid flows; apparatus 2 – 2,5-5 m^3/hour; apparatus 3 – 5-10 m^3/hour. Proposed construction provides anti-agglomerator synthesis carrying out in a wide diapason of process productivity 2,5-10 m^3/hour. Application of novel technology at stage of interaction of potassium stearate with calcium chloride allow expecting the following: stoichiometric reaction proceeding, decrease of reaction mixture residence time in apparatus, receipt of suspension of calcium stearate of high quality (firstly by dispersion), simplicity of reactor exploitation, increase of apparatus exploitation time between shutoffs due to drivings by reaction products..

LOW-MOLECULAR PRODUCTS PRODUCTION IN THE FLOW

Unblended Benzene Production

It was shown in chapter 3 that flows of liquid components differ in density and viscosity at least in 20%, in particular infinitely mixing glycerin (ρ = 1,26 g/cm^3) and water (ρ = 1,0 g/cm^3), are segregated in tank apparatus. So, creation of homogeneous phase is a problem.

On the plant production of unblended benzine by combination of 7 components differ in density (technology of turbulent compounding) was organized. Non-optimal conditions of mixing didn't allow continuous production of benzine of required quality. Application of multizone tubular turbulent apparatus of divergent-convergent design (Figure 10) solved the problem.

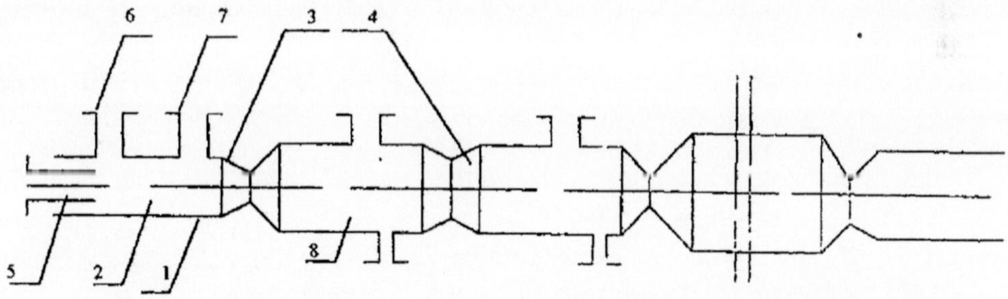

Figure 10. Tubular turbulent apparatus for preparation of unblended benzine (1-8 – see description text).

Device has body 1 made as thick-walled cylinder in which static mixer elements in the form of combined cones are placed – convergents 3 and divergents 4, they divide the device on sections. Consecutive introduction of components and mixture mixing regime are reached at the expense of apparatus constructive particularities and allow effective compounding of fractions with various boiling temperature and with large interval of volume throughput that is of great importance when compositions of motor petrol are prepared in production quantities. Mixing of high-boiling components of fuel composition in the first section allows receipt of stable components mixture and then in subsequent sections to mix it with the rest components flows, for example with virgin benzin, benzine of catalytic cracking or refining, and by this stable in structure unblended benzine is continuously formed.

Alkyl-Sulfates Production

Interaction of olefine hydrocarbons with sulfuric acid is widely used in industrial production, in particular under receipt of synthetic cleaning agents, methylethylketone, benzine fractions of hydrocarbons [6], higher fatty alcohols [7], etc.:

$$RCH{=}CH_2 + H_2SO_4 \longrightarrow R\text{-}\underset{\displaystyle CH_3}{\underset{|}{C}H}\text{-}O\text{-}SO_3H + Q$$

Anionic surfactant species on the base of alkylsulfates are widely used as a base for synthetic cleaning agents, emulsifiers, stabilizers of foams, etc.. Alkylsulfates $ROSO_3Me$ (Me – Na, K and so on) at conditions of industrial production, in particular under synthesis of cleaning agents are produced under interaction of α-olefines fractions of unsatureated hydrocarbons C_8-C_{18} (olefins content not lower than 97 mass %) with 98% sulfuric acid at the ratio α-olefine / H_2SO_4 = 1/1,2 ÷ 1,4 mole with further neutralization of unreacted H_2SO_4 and saponification of received alkyl-sulfuric acid:

$$R\text{-}\underset{\displaystyle CH_3}{\underset{|}{C}H}\text{-}O\text{-}SO_3H + NaOH \longrightarrow R\text{-}\underset{\displaystyle CH_3}{\underset{|}{C}H}\text{-}O\text{-}SO_3Na + H_2O$$

Traditional technological scheme of alkylsulfates receipt includes following stages (Figure 11a) [8]: sulfation of α-olefines by concentrated sulfuric acid (position 1a, 1b); neutralization of unreacted H_2SO_4 and alkylsulfuric acid (position 2a-2d); two-stage cleaning by benzine of received alkylsulfates from unsaponifiable hydrocarbons products (position 4a, 4b, 5).

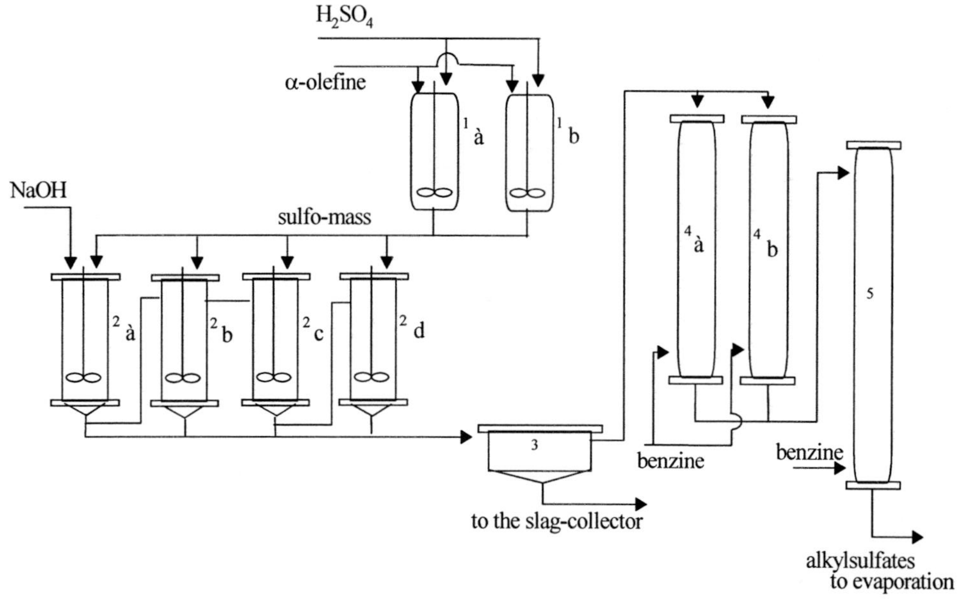

Figure 11. Continued

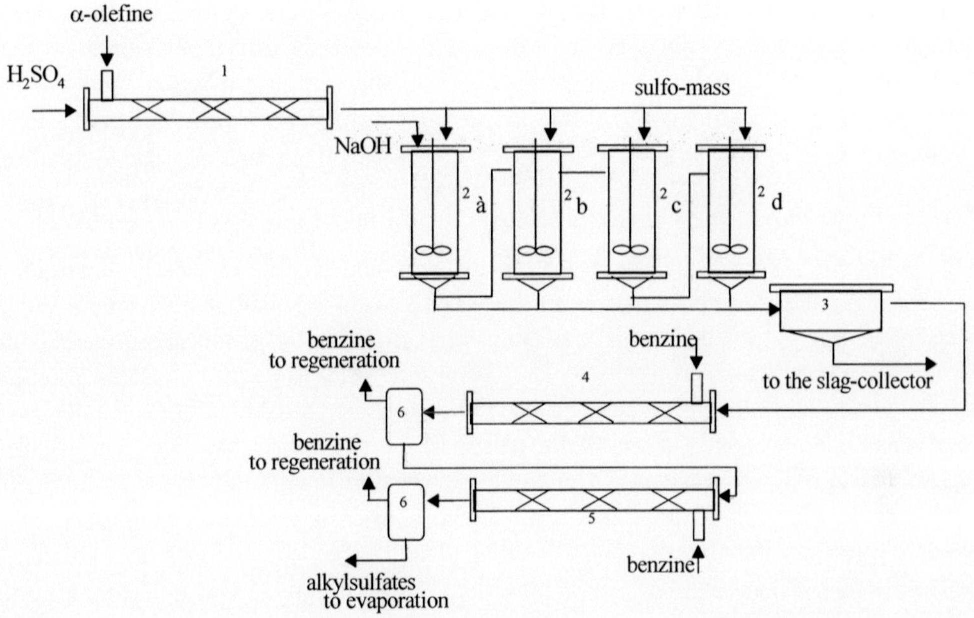

Figure 11. Technological scheme for receipt of alkylsulfates: 1a, 1b – tank sulfators with impeller mixers (v_r = 0,157-0,44 m^3); 4a, 4b – extraction columns of the first stage (v_r = 3,4 m^3); 5 – extraction column of the second stage (v_r = 4,7 m^3) (scheme a), 1 – tubular turbulent sulfator (v_r = 0,0086 m^3); 4 – tubular turbulent extractor of the first stage (v_r = 0,003 m^3); 5 – tubular turbulent extractor of the second stage (v_r = 0,005 m^3); 6 – sediment box (scheme b), 2a-2d – tank reactors-neutralizator with propeller mixers (v_r = 3 m^3);. 3 – seoarator.

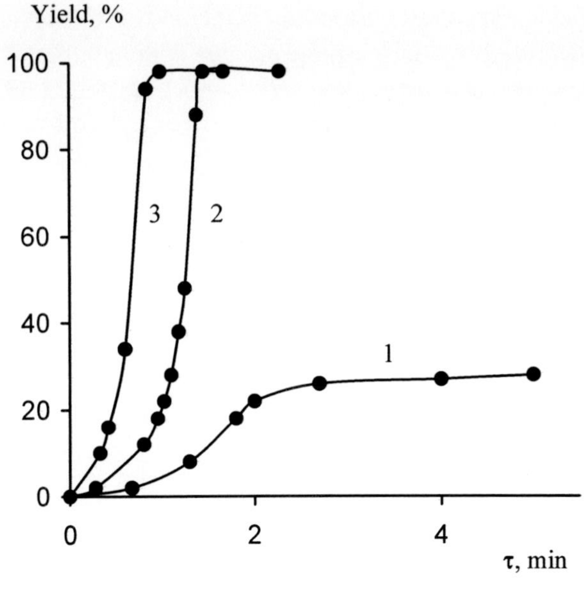

Figure 12. Conversion curves of sulfation of α-olefines of fraction C_8-C_{16} by concentrated (95 mass %) sulfuric acid in stirred tank reactors. Ratio α-olefines / H_2SO_4 = 3,9 volume. 298 K. Mixing rate, revs: 100 (1), 500 (2), 1000 (3).

Selection of reactor for sulfation of α-olefines caused significant complications due to specificity of chemical process proceeding. Reaction proceeds fast enough (characteristic reaction time according with [7] is lower than 30 sec) and is high-exothermal. In particular, heat of sulfation is equal to $q \approx 502{,}1 \cdot 10^3$ kJ per 1 m^3 of α-olefines [8]. As a consequence temperature rise in reaction zone is about 190 ± 5^0C. We should also note that due to significant difference in values of viscosity (μ) and density (ρ) of initial reagents ($\mu_{H_2SO_4}/\mu_{olefine} \approx 75{,}1$; $\rho_{H_2SO_4}/\rho_{olefine} \approx 2{,}63$) the flows are mixed badly and by this especially at the initial period reaction proceeds on phases contact surface that leads to deceleration of chemical reaction proceeding in spire of the fact that this reaction may be ascribed to fast chemical processes. In this case increase of turbulent level of mixing in reaction zone leads to the rise of chemical process proceeding rate (Figure 12). As a consequence under technological design of alkylsulfuric acid receipt stage in production of alkylsulfates it is necessary to form in reaction zone intensive turbulent mixing of initial reagents and to provide removal of resulted in the course of reaction heat. Moreover, presence of side processes (formation of dialkylsulfates, resinification of α-olefines, hydrolysis, and also alcoholysis of dialkylsulfates, etc.) determines the necessity of reaction mixture residence times τ_r reduction and definite limitation of upper temperature level in reaction zone (reactor).

This required intensive turbulent mixing of liquid flows in apparatus, large specific surface for improving of external heat removal, jet regime under reaction with easy controlled reagents residence time τ_r, that are possible under tha use of apparatus of novel generation – compact tubular turbulent reactors for effective carrying out of reaction of α-olefines sulfation by sulfuric acid at conditions of industrial production.

One of the real ways of essential decreasing of adiabatic temperature rise in reaction zone ΔT_{ad} under α-olefines sulfation by sulfuric acid is dilution of initial reaction mixture by inert solvent (saturated hydrocarbons, for example decane) with high heat capacity (Table 3) and there is also significant reduction of cooling zone length L_{cool} (Figure 13). Application of low-boiling solvent for dilution of initial reaction mixture allows effective use also of internal heat removal at the expense of its boiling. In this case gasification in reaction mixture volume allows additionally turbulization of flow and high linear rates of movement prevent from gas "locks" formation. The process becomes simple, compact, effective and easy control.

As a result of proceeding of α-olefines sulfation reaction by concentrated sulfuric acid in reaction volume heat Δq_{ad} is evaluated:

$$\Delta q_{ad} = q w_{olefine} \qquad (1)$$

Here $w_{olefine}$ – volume throughput of hydrocarbon fraction of α-olefines.

The heat q_{ad} is spent on adiabatic heating of reaction mass up to the temperature determined by solvent boiling temperature T_{boil}:

$$q_{ad} = q w_\Sigma \rho (T_{boil} - T_0) \qquad (2)$$

and using equations (1) and (2) one may obtain amount of low-boiling solvent w_s required for removal of the residuary heat amount at the expense of boiling:

$$w_s = \frac{\Delta q_{ad} - q_{ad}}{q_{ev}} \quad (3)$$

where q_{ev} – solvent evaporation heat.

Table 3. The influence of dilution of α-olefines by decane on heat regime of sulfation in tubular turbulent apparatus without internal heat removal (T_0 = 253 K, T_{ch} = 253 K, R = 0,014 m, V = 0,35 m/sec, apparatus length 14 m)

Components consumption, m^3/hour			Ratio α-olefine/ decane, volume	ΔT_{ad}, ^{0}C	Temperature in reaction zone, K	
α-olefine (fraction C_8-C_{18})	decane	H_2SO_4 (98 mass %)			T_{ad}	T_r
0,62	-	0,160	-	189	442	385
0,34	0,34	0,088	1:1	112	365	326
0,24	0,48	0,062	1:2	81	334	303
0,18	0,55	0,470	1:3	64	317	290
0,15	0,59	0,038	1:4	53	306	283
0,12	0,62	0,032	1:5	46	299	277
0,11	0,64	0,031	1:6	39	292	273,6

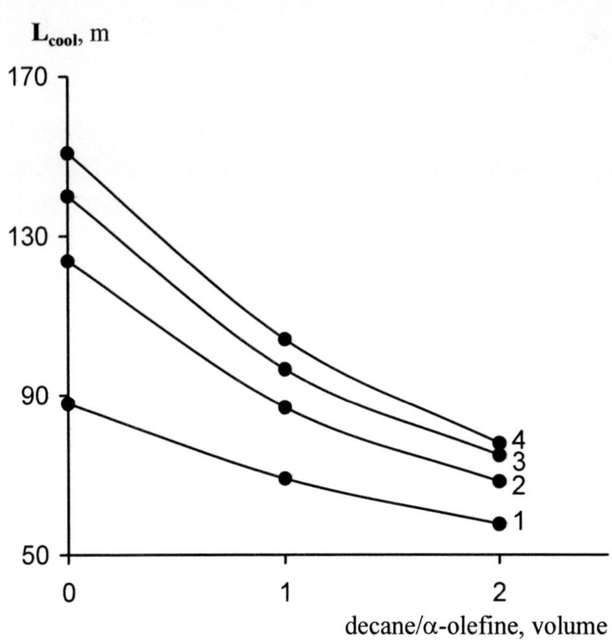

Figure 13. Dependence of the length of cooling zone L_{cool} on dilution of initial reaction mixture under the use of "zone" model. Introduction of equal amounts of H_2SO_4 in 1 (1), 2 (2), 3 (3) and 4 (4) reaction zones at the absence of internal heat removal (T_0 = 303 K, T_s = 278 K, T_{ch} = 253 K, R = 0,014 m, $w_{olefine}$ = 0,62 m^3/hour, w = 0,16 m^3/hour).

For example, under dilution of initial reaction mixture by petroleum ether with $T_{boil} \approx 343$ K in ratio α-olifine / solvent = 1 / 1,24 volume at the expense of adiabatic heating of reacting mixture from $T_0 = 293$ K up to $T_{boil} = 343$ K ($\Delta T = 50^0C$) in tubular turbulent apparatus at $w_{olefine} = 0,62$ m^3/hour, $W_{H_2SO_4} = 0,16$ m^3/hour and R = 0,014 m $1,5 \cdot 10^5$ kJ of heat are removed, i.e. about 48% of heat. At the expense of solvent boiling (petroleum ether) at atmosphere pressure in apparatus $1,6 \cdot 10^5$ kJ are removed, i.e. residuary 52% of heat. If one will use dilution in ratio α-olefine / petroleum ether = 1 / 1 volume, so at the expense of adiabatic temperature rise and solvent boiling about 86% of heat will be removed and temperature in sulfator will rise at the expense of auto-thermal heating in 15-17°C higher than boiling of solvent that is permissible. In the last case after removal of the main petroleum ether part out of gas phase in real production the scheme of alkyl-sulfuric acid purification from solvent's remains is simplified due the absence of the last one. The whole technological scheme is also simplified on the base of compact tubular turbulent apparatus (Figure 6.11b) in comparison with the scheme existing in production now (Figure 11a).

There is also alternative method of effective decrease of adiabatic temperature rise at points of reagents introduction – "zone" model of process carrying out (Figure 11d) at the expense of distribution of given amount of sulfuric acid along the length of tubular turbulent apparatus of cylindrical design. In particular increase of number of zones of portion introduction of lower amount of H_2SO_4 at introduction points allows sharp reduction of ΔT_{ad} and preventing of sharp temperature rise in reaction zone higher than technologically acceptable value (Figure 14).

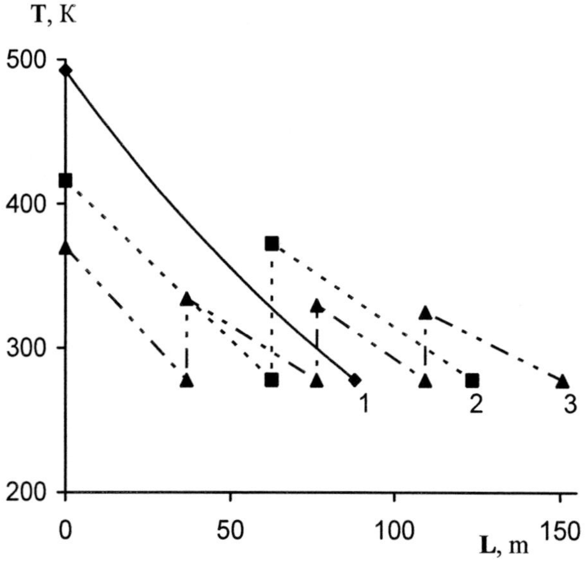

Figure 14. Temperature profile T along tubular turbulent sulfator length L under the use of "zone" model. Introduction of equal amounts of H_2SO_4 into 1 (1), 2 (2), and 4 (3) reaction zones (condition as in Figure 13).

As a consequence, under increase of number of reagents introduction zones in tubular apparatus "deceleration" of fast reaction is observed expressed in the fact that tempetature profile in reactor approaching to chemical process proceeding at $\tau_{ch} \approx \tau_{pr}$ in ideal mixing regime. In this case even at the absence of external heat removal reaction mixture temperature over the whole apparatus length is equal, i.e. process proceeds in quasi-isothermal regime.

However, under the use of "zone" model of process carrying out the technologic scheme becomes complicated, in particular at the expense of necessity of use of several cooling devices between reaction zones and increase of summary length of cooling zone L_{cool} and consequently reactor's dimensions (Figure 14).

For simplification of process' technological line under the use if zone model, in particular under sulfation of α-olefines by sulfuric acid at the cooling stage it is advisable to use shell-and-tube turbulent apparatus of jet type including pipe bundle of small diameter washing by coolant (Figure 1c). Fragmentation of flow into N tubes of smaller radius r and working in quasi-plug flow regime at maintaining of general section of flow S at in single tubular apparatus under the use of both traditional and zone models allows reduction of cooling zone length L_{cool} in $N^{0,6}$ times and making the process more compact. Combination of portion introduction of sulfuric acid and application of multi-tubing cooling inserts allow additional decreasing of temperature adiabatic rise ΔT_{ad} and significant reduction of cooling zone length under sulfation of α-olefines (Figure 15).

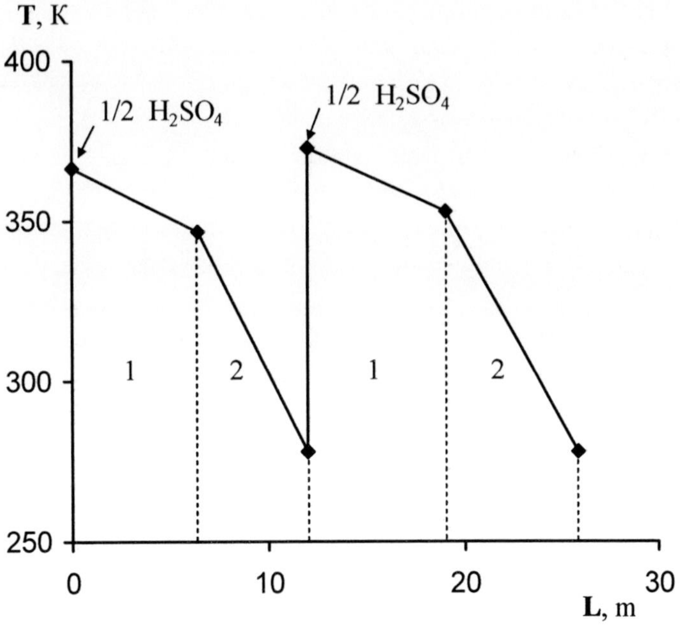

Figure 15. Temperature profile along tubular turbulent sulfator length under two-stage introduction of equal amounts of H_2SO_4. Reaction zone (1) – tubular turbulent apparatus of cylindrical design (R = 0,014 m); cooling zone (2) – shell-and-tube turbulent apparatus (N = 31, r = 0,0025 m). T_0 = 253 K, T_s = 278 K, T_{ch} = 253 K, $W_{olefine}$ = 0,62 m^3/hour, $W_{H_2SO_4}$ = 0,16·m^3/hour).

Thus, reaction of α-olefines sulfation by concentrated sulfuric acid is related to fast chemical reactions that are necessary to carry out in compact energy- and resource-saving tubular turbulent apparatus. For guarantee prevention of sharp temperature rise in reaction zone under sulfation of α-olefines application of tubular turbulent apparatus of cylindrical and shell-and tubes designs is advisable. By this one should dilute initial reagents especially by low-boiling solvent and realize "zone" model of process carrying out (portion reagents introduction along reactor length). Novel energy- and resource-saving technological scheme of alkylsulfates receipt at conditions of industrial production with the use of compact tubular turbulent apparatus at the stage of α-olefines sulfation by sulfuric acid and washing of resulted alkylsulfates was proposed.

Neutralization of Acid and Alkali Mediums

Stage of neutralization of acid mediums in liquid-phase flows is integral part of the majority of chemical processes of synthetic products production in industry including polymers synthesis. As the damping of industrial wastewaters to sewerage system is permitted at $pH \geq 7$, then stage of neutralization of acid mediums is the key stage under purification of industrial effluents and also under receipt of many synthetic products (neutralization of aminochlorohydrates under ehtylenediamine production, ammonation of phosphoric acid under production of ammophos, neutralization and saponification of alkylsulfuric acid under alkylsulfates production, washing out of inverse solvent in polymers production, etc.). Under neutralization stage carrying out at conditions of industrial production it is necessary to take into account specific particularities of this process. Neutralization reaction is very fast (length of reaction zone is very small $L_{ch} \approx 2 \cdot 10^{-10}$ m). This caused its proceeding in classic stirred tank reactors in diffusion regime even at low reagents concentrations, by this neutralization reaction is highly exothermal process (heat effect of interaction of strong acid with strong base is $q = 57{,}2$ kJ/mole). Process proceeds in aggressive mediums that at high temperature require anti-corrosion apparatus design. Under process optimization effective mixing of initial reagents solvents for the time τ_{mix} commensurable with the time of chemical reaction τ_{ch}, i.e. at $\tau_{mix} \approx \tau_{ch}$ and also application of apparatus without moving mechanical devices with small specific area of reaction mixture with reactor's walls contact and decreasing of sharp temperature rise in neutralization agent introduction zone are required.

Experience of application of apparatus of novel generation – compact tubular reactors functioning in regime of quasi-plug flow mode in turbulent flows without application of mixers for fast and many of mass-exchange physical processes in combination with results of laboratory research and mathematical modelling of liquid flows mixing processes allow us to suggest their effective functioning under neutralization of acid effluents that can not be reached under the use of any other known apparatus of applied chemistry. In this case it is important to reveal possibilities of formation of plan reaction front in reaction zone corresponding to quasi-plug flow mode in turbulent flows and providing quasi-isothermal regime in reaction zone.

Obtained experimental results allow derivation of dependence of conditions of quasi-plug flow mode formation in tubular turbulent apparatus at V_1, $V_2 = 0{,}1 \div 0{,}8$ m/sec and $d_1 / d_d = 0{,}44$:

$$V_1/V_2 = 0{,}28 \cdot V_2 + 0{,}96 \qquad (4)$$

One can estimate the fact that under neutralization of 30% water solution of HCl by NaOH water solution temperature rise ΔT_{ad} is equal to 50°C, at the same time under neutralization of 30% acid solution by solid alkali ΔT_{ad} will be significantly higher and about 130°C (Table 4).

Table 4. Temperature rise in the course of neutralization of water solution of hydrochloric acid by sodium hydroxide

C_{HCl}, mass %	C_{NaOH}, mass %					
	10		20		100	
	C_{NaCl}, mole/l	ΔT_{ad}	C_{NaCl}, mole/l	ΔT_{ad}	C_{NaCl}, mole/l	ΔT_{ad}
0,01	0,0027	0,04	0,0027	0,04	0,0027	0,04
0,02	0,0055	0,08	0,0055	0,08	0,0055	0,08
0,05	0,0136	0,19	0,0137	0,19	0,0137	0,19
1	0,2500	3,4	0,2699	3,7	0,2748	4
5	0,9169	12	1,1185	15	1,3699	19
10	1,4105	19	1,9528	27	2,8712	39
30	2,1433	29	3,7043	51	9,4438	129

It is obvious from presented results that process of neutralization of acid mediums (to the extent of 30% acid water solution) may be carried out by water solution of alkali (up to 20 mass %) without additional external heat removal at initial temperature of acid effluent not higher than 30°C. in this case temperature rise ΔT_{ad} in reaction zone will not exceed 30-50°C (Table 4).

Under the use of reagents concentrated solutions for neutralization of acid mediums sharp temperature rise is observed. In this case the novel type of apparatus of applied chemistry – compact tubular turbulent reactors of cylindrical or divergent-convergent designs determine the possibility of effective control of temperature field in reaction zone by several ways: change of apparatus radius and reagents flow rate, application of "zone" model for fast chemical process carrying out and the use of shell-and-tube apparatus with pipe bundle of N tubes of small radiuses.

It should be noted that in the necessity of effective mixing of reagents and preventing of possible acid slip caused by the fact that real flow of industrial effluent significantly exceeds the consumption of neutralization agent it is advisable to carry out the reaction in tubular turbulent apparatus of divergent-convergent design. In this case one should apply reactor with diameter in the wide part (divergent) from 0,05 up to 0,4 m and length up to 2 m (apparatus volume is about 0,016-0,12 m^3) that corresponds the productivity of one compact reactor about 7 m^3/hour – 460 m^3/hour and higher. The holes in socket for alkali (acid) solution introduction into reactor (placed coaxialy to the flow) should provide reagents introduction into reaction zone with the rate $V_1 > 1,1$ m/sec to provide the ratio $V_1 / V_2 \geq 1,2$.

Thus, reaction of neutralization of acid and alkali mediums relates to fast chemical processes which are advisable to carry out in reactors of novel type – compact energy- and resource-saving tubular turbulent apparatus of cylindrical or divergent-convergent designs that provides jet regime of reagents flow, quasi-isothermal conditions in reaction zone, total conversion of reagents, etc.. Neutralization of industrial effluent with mass content of acid up to 30 mass % can be realized in tubular turbulent apparatus by both 10% and 20% water alkali solution without additional heat removal. Results obtained in this work may be also used under realization of neutralization of alkali effluent by acids.

REFERENCES

[1] K.S. Minsker, Al.Al. Berlin, V.P. Zakharov, G.E. Zaikov. *Fast liquid-phase processes in turbulent flows*. Brill Ac. Publ. VSP. Netherlands. 2004.180 p.

[2] Al. Al. Berlin, K. S. Minsker, K. M. Dyumaev *New Unified Energy- and Material-Saving, High-Performance, Ecological-Friendly Technologies based on Tubular Turbulent Reactors*, Moscow (1996) *(in Russian)*.

[3] K. S. Minsker, Al. Al. Berlin, V. P. Zakharov *Vysokomol. Soedin.*, 44, No. 9, 1606-1627 (2002).

[4] P. A. Kirpichnikov, V. V. Beresnev, L. M. Popova *Album of Technological Schemes of the Base Productions in Synthetic Rubber Industry,* Khimiya, Leningrad (1986) *(in Russian)*.

[5] P. A. Kirpichnikov, L. A. Averko-Antonovich, Yu. O. Averko-Antonovich *Chemistry and Technology of Synthetic Rubber,* Khimiya, Leningrad (1975) *(in Russian)*.

[6] Al. Al. Berlin, K. S. Minsker, K. M. Dyumaev *New Unified Energy- and Material-Saving, High-Performance, Ecological-Friendly Technologies based on Tubular Turbulent Reactors*, Moscow (1996) *(in Russian)*.

[7] S. M. Loktev, V. L. Klimenko, V. V. Kamozoskin, A. T. Menyailo *The Higher Butters,* Khimiya, Moscow (1970) *(in Russian)*.

[8] B. R. Serebryakov, T. K. Plaskunov, V. R. Nizheles, M. A. Dalin *The Highest Olefines,* Khimiya, Moscow (1984) *(in Russian)*.

INDEX

A

B

D

E

F

J

K

L

M

T

U

V

W

X

Y